JAMES K. CLARKE

MATHEMATICS APPLICATIONS
UNIT 1

Copyright © 2015 JAMES K. CLARKE

Published by Vivid Publishing
P.O. Box 948, Fremantle
Western Australia 6959
www.vividpublishing.com.au

Mathematics Applications Unit 1 - 1st edition

Cover designed by Nicholas Miles-Gray
Answers checked by Mitchell Ford, Travis Williams
Project edited by Isaac Maldonado, Deepesh Gendah

National Library of Australia Cataloguing-in-Publication data:
Creator: Clarke, James K., author.
Title: Mathematics applications Unit 1. Book 1 / James K. Clarke.
ISBN: 9781925209747 (paperback)
Target Audience: For secondary school age.
Subjects: Mathematics--Study and teaching (Secondary)--Australia.
            Mathematics--Problems, exercises, etc.
Dewey Number: 510.76

All rights reserved. No part of this publication may be reproduced, stored in a retrieval system or transmitted in any form or by any means, electronic, mechanical, photocopying, recording or otherwise, without the prior written permission of the copyright holder. The information, views, opinions and visuals expressed in this publication are solely those of the author(s) and do not necessarily reflect those of the publisher. The publisher disclaims any liabilities or responsibilities whatsoever for any damages, libel or liabilities arising directly or indirectly from the contents of this publication.

PREFACE

Mathematics Applications Unit 1 is a course-book based designed for students who want to prepare for the Western Australian Certificate of Education (WACE) examination. The scope of the book is to extend the mathematical skills of students beyond Year 10 level.

Mathematics Applications **Unit 1** is a course-book which focuses on the use of mathematics to solve problems in contexts that involve mathematical modelling. The book covers three main topics which are: **Consumer Arithmetic**, **Algebra and Matrices** and **Shape and Measurement**. The latter have been subdivided into ten chapters as shown in the table of contents.

Consumer Arithmetic reviews the concepts of rate and percentage change in the context of earning and managing money, and provides a context for the use of spread sheets.

Algebra and Matrices, on the other hand, continues the Year 7–10 study of algebra and introduces the new topic of matrices. The emphasis of this topic is the symbolic representation and manipulation of information from real-life contexts using algebra and matrices.

Shape and Measurement extends the knowledge and skills students developed in the Year 7–10 curriculum with the concept of similarity and associated calculations involving simple and compound geometric shapes. The emphasis in this topic is on applying these skills in a range of practical contexts, including those involving three-dimensional shapes.

Mathematics Applications Unit 1 is a book that includes notes, formulae, guidance of how to make use of technology to solve problems and worked examples that follow a simple, logical and easy-to-understand approach. The exercises have been carefully constructed to reinforce the students' learning of concepts and skills in Mathematics taught in the classroom. The questions are interesting, stimulating and challenging, and will help students develop skills and understanding necessary for them to excel in Mathematics.

Worked solutions have been included for each topic, at the end of the book, to assist in the teaching and learning of mathematical skills.

TABLE OF CONTENTS

1. Use of formulae — page 7
1A	Evaluating linear and non-linear expressions	page 7
1B	Determining the value of the subject of a formula	9
1C	Use of formulae and Technology (CAS)	10
1D	Applications	13
1E	Spreadsheets	17

2. Percentages — page 19
2A	Expressing one quantity as a percentage of another	20
2B	Finding a given percentage of an amount	22
2C	Increase and decrease in percentages	23
2D	Calculating an original amount	25
2E	Profit and loss	26
2F	Discount	29
2G	Commission	31
2H	Calculating tax paid and taxable income	34

3. Simple Interest — page 37
3A	Calculating Simple Interest	37
3B	Simple Interest and Technology	40
3C	Finding Principal, Rate of Interest or Time	41

4. Compound Interest — page 43
4A	Calculating Compound Interest	43
4B	Compound Interest and Technology	45
4C	How much do I repay	48
4D	Inflation	49
4E	Depreciation	51

5. Financial Mathematics — page 53
5A	Calculating wages from an annual salary	53
5B	Calculating wages from an hourly rate	55
5C	Overtime and other allowances	57
5D	Piecework	59
5E	Exchange rates	61
5F	Best buys	64
5G	Budgeting	65
5H	Shares and dividends	69

	5I	Price-to-Earnings ratio	72
	5J	Allowances : Family Tax Benefit	74
	5K	Carer's Allowance and Old age pensions	76

6. Matrices   page 79

	6A	Size (order) of a matrix	80
	6B	Types of matrices	81
	6C	Addition of matrices	83
	6D	Subtraction of matrices	84
	6E	Multiplying a matrix by a scalar	85
	6F	Multiplication of matrices	87
	6G	Determining the power of a matrix using technology	90
	6H	One stage Adjacency matrix	91
	6I	Two stage adjacency matrix	93
	6J	Social Interaction as a matrix	96
	6K	Storing, displaying information, model and solve problems	99

7. The Theorem of Pythagoras   page 103

	7A	How to prove a triangle is right angled?	103
	7B	Mental skills	105
	7C	Finding the length of the hypotenuse	107
	7D	Finding the length of the shorter sides	109
	7E	Applications in two dimensions	111
	7F	Applications in three dimensions	114

8. Perimeter and Area   page 117

	8A	Perimeter	117
	8B	Circumference of a circle	119
	8C	Arc length and perimeter of a sector	120
	8D	Converting area units	122
	8E	Area formulae	123
	8F	Finding missing length and radius	126
	8G	Applications	129

9. Surface Area and Volume   page 133

	9A	Surface area and Volume formulae	133
	9B	Finding radius or side	140
	9C	Capacity	144
	9D	Volume of other prisms	146
	9E	Applications	147

10. Similarity — page 151

10A	Similar figures and similar triangles	151
10B	Testing similar triangles	152
10C	Similarity and scale factor	155
10D	Area of similar figures	159
10E	Volume of similar figures	162
10F	Maps, Scale drawings and building plans	166

WORKED SOLUTIONS — **169+**

CHAPTER 1

USE OF FORMULAE

1A EVALUATING LINEAR AND NON-LINEAR EXPRESSIONS

Consider the formula $T = 2\pi\sqrt{\dfrac{l}{g}}$, where T is the time period for one oscillation of a pendulum, l is the length of the pendulum and g is the acceleration due to gravity.

In mathematics, the formula $T = 2\pi\sqrt{\dfrac{l}{g}}$ is an entity constructed using the symbols. For example, to determine the time period for one oscillation (T) we need to know the length of the pendulum and the value of g in question. We can quickly and easily determine the value of T. Note that the time T and the length l are expressed as single letters instead of words or phrases.

In this chapter, the reader will be expected to substitute numerical values into algebraic expressions, and evaluate ; both linear and non-linear expressions. With complicated numerical manipulation the use of technology (CAS) is expected.

EXAMPLES

1. Given that $a = -2$ and $b = 3$, find the value of (i) $3a + 2b$ (ii) $a^2 + b^2$ **SOLUTION** Replace the value of a by -2 and b by 3. It is of utmost importance to always use brackets to avoid unpleasant surprises. (i) $3(-2) + 2(3) = -6 + 6 = 0$ (ii) $a^2 + b^2 = (-2)^2 + (3)^2$ $ = 4 + 9 = 13$	2. Given that $p = 4$ and $q = -3$, evaluate (i) pq^2 (ii) $(2p + q)^3$ **SOLUTION** (i) $4(-3)^2 = 4 \times 9 = 36$ (ii) $[2 \times 4 + (-3)]^3 = (8 - 3)^3$ $ = 5^3$ $ = 125$
3. Given that $a = 11$ and $b = -3$, find the value of (i) $\sqrt{2a - b}$ $\sqrt{2(11) - (-3)} = \sqrt{25} = 5$ (ii) $\sqrt[3]{2a - b + 2}$ $\sqrt[3]{2(11) - (-3) + 2} = \sqrt[3]{27} = 3$	4. Given that $x = 3$, $y = 4$ and $z = -2$, evaluate (i) $2x^2 - y$ $2(3)^2 - 4 = 18 - 4 = 14$ (ii) $y(x - z)$ $4(3 - (-2)) = 20$

EXERCISE 1A

1. Given that $a = 4$ and $b = -3$, find the value of

   (i) $2a + 5b$

   (ii) $7 - b^2$

2. Given that $p = 3$ and $q = -5$, evaluate

   (i) $p + q^2$

   (ii) $(3p + 2q)^2$

3. Given that $a = -5$ and $b = 2$ and $c = 0$, find the value of

   (i) abc

   (ii) $a - 2b + 3c$

   (iii) $a^2 + b^2 - c^2$

4. Given that $p = -2$ and $q = 7$, evaluate

   (i) $4p + 3q$

   (ii) $3p^2$

   (iii) $5q - 4p$

5. Given that $a = 5$ and $b = -2$, find the value of

   (i) $(a - b)^2$

   (ii) $a^3 - b^3$

   (iii) a^b

6. Given that $p = 2$, $q = -3$ and $r = 4$, evaluate

   (i) $4p + q - 2r$

   (ii) $5p^2$

   (iii) $(p + 2r)(q + 2r)$

1B DETERMINING THE VALUE OF THE SUBJECT OF A FORMULA

At this stage of the chapter, we have to determine the value of the subject of a formula given the values of the other pronumerals in the formula. (Transposition not required)

EXAMPLE 1	EXAMPLE 2
Given the formula $y = mx + c$, find y when $m = 4, x = -2$ and $c = 11$. **Solution** $y = 4(-2) + 11 = 3$	Given that $v^2 = u^2 + 2as$, find the values of v when $u = 3, a = 10$ and $s = 3.6$. **Solution** $v^2 = 3^2 + 2(10)(3.6) = 81$ $\therefore v = \pm 9$

EXERCISE 1B

1. Given the formula $v = u + at$, find v when $u = 25$, $a = 10$ and $t = 3$.	2. Given that $s = ut + \frac{1}{2}at^2$, find the value of s when $u = 20$, $a = 9.8$ and $t = 6$.
3. Given the formula $A = 2\pi r^2 + 2\pi rh$, find A when $r = 4$ and $h = 9$. (Use $\pi = 3.14$)	4. Given that $V = \frac{4}{3}\pi r^3$, find the value of V when $r = 5$.
5. The sum (S) of positive integers from 1 to n is given by $S = \frac{1}{2}n(n+1)$. Find S when $n = 12$.	6. Given that the area (A) of a triangle is given by $A = \frac{ab \sin C}{2}$, find the value of A when $a = 5$, $b = 10$ and $C = 30°$.
7. Einstein's famous equation relating energy (E), mass (m) and speed of light (c) is $E = mc^2$. Find E when $m = 0.0001$ and $c = 3 \times 10^8$.	8. Given that $y = a + bx^2$, find the value of y when $a = 7, b = 4$ and $x = -5$.

1C USE OF FORMULAE AND TECHNOLOGY (CAS)

Where the numerical manipulation is complicated, the use of technology is really helpful. Use the following steps on your CAS:

- Menu
- Num Solve
- Insert your formula
- Input given values
- Solve

EXAMPLES

Use the solve facility on your calculator to attempt the following questions.

1. Given that $v^2 = u^2 + 2as$, find the value of v ($v > 0$) when $u = 5$, $a = 6$ and $s = 50$.

Remember to click the bubble at v as we are solving for v.

$$v = 25$$

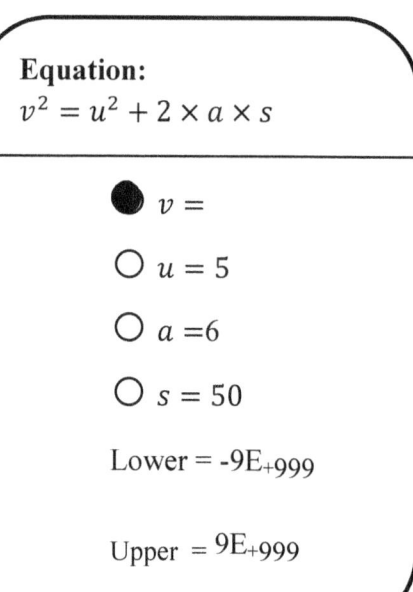

2. Given that $c = b^2(a + b) + b^3 - a^2$, find c when $a = 3$ and $b = 5$.

SOLUTION

$c = 316$

Equation:
$c = b^2 \times (a + b) + b^3 - a^2$

● $c =$
○ $a = 3$
○ $b = 5$

Lower = -9E+999

Upper = 9E+999

EXERCISE 1C

Use Num Solve to find the missing pronumerals in each case.

1. $V = \frac{1}{3}\pi r^2 h$ Evaluate V (a) when $r = 6$ and $h = 12$. (b) when $r = 5$ and $h = 10$. (c) when $r = 8$ and $h = 52$.	2. $s = \frac{1}{2}(u+v)t$, find the value of s (a) when $u = 30, v = 50$ and $t = 2$. (b) when $u = 40, v = 60$ and $t = 3$. (c) when $u = 20, v = 48$ and $t = 3$.
3. The formula for the area of a trapezium is $A = \frac{1}{2}h(c+d)$. Evaluate A when (a) $h = 8$, $c = 11$ and $d = 13$. (b) $h = 16$, $c = 15$ and $d = 25$.	4. Find the value of $A = \frac{a+\sqrt{a^2+b^2}}{a^2-2ab}$ when $a = -4$ and $b = -3$. Give your answer as a fraction.
5. $\frac{1}{b} = \frac{1}{c} + \frac{1}{d}$, Evaluate b when c = 3 and d = 8.	6. $\frac{1}{x} = \frac{1}{y} + \frac{1}{z}$, Evaluate x when y = 4 and z = 10.
7. Given $v = u + at$, find v given that (a) $u = 10$, $a = 5$ and $t = 7$. (b) $u = 55$, $a = 10$ and $t = 4$.	8. If $E = \frac{1}{2}mv^2$, find the value of E when (a) $m = 100$ and $v = 5$. (b) $m = 200$ and $v = 25$.

9. In the formula $y = mx + c$, find y given that
  (a) $m = 5, x = 10$ and $c = 3$.

  (b) $m = -5, x = 8$ and $c = 11$.

  (c) $m = 0.5, x = 12$ and $c = 17$.

10. In the formula $E = mc^2$, find E given that
  (a) $m = 50$ and $c = 5$.

  (b) $m = 60$ and $c = 8$.

11. Given that $v^2 = u^2 + 2as$, find the value of v when
  (a) $u = 20, a = 8$ and $s = 50$.

  (b) $u = 30, a = 10$ and $s = 100$.

12. Given that $A = \frac{a+b}{2} \times h$, find the value of A when $a = 18, b = 12$ and $h = 9$.

13. $V = \pi r^2 h$, find the value of V given that
  (a) $r = 8$ and $h = 12$.

  (b) $r = 7$ and $h = 20$.

14. $V = 2\pi r^2 + 2\pi r h$, find the value of V given that $r = 6$ and $h = 11$.

CHAPTER 1 : USE OF FORMULAE

1D   APPLICATIONS

1. Body Mass Index (BMI) is an index of weight-for-height that is commonly used to classify underweight, overweight and obesity in adults. It is defined as the weight (W) in kilograms divided by the square of the height (H) in metres (kg/m²). Thus,
$$BMI = \frac{W}{H^2}$$

   (a) Use the above formula to determine the BMI of an adult who weighs 70kg and whose height is 1.75m.

   (b) A BMI of 25 or more is classified as over-weight. Sarah weighs 80kg and has a height of 1.65m. Is she overweight? Support your answer by calculations.

   (c) A BMI of less than 18.5 is normally considered as underweight. Peppa is 51kg and has a height of 150 cm. Calculate Peppa's BMI to see whether he is under weight or not.

2. The formula used to convert Fahrenheit (F) to Celsius (C) is given by $C = \frac{5}{9}(F - 32)$. Use the formula to convert

   (a) 48°F into °C.

   (b) 108°F into °C.

   (c) 49°C into °F.

3. The Kinetic Energy (E) is the energy possessed by a body because of its motion and is equal to half the mass (m) of the body times the square of its speed (v).
Symbolically, $E = \frac{1}{2}mv^2$.

(a) Calculate the kinetic energy of a particle having mass 50kg and speed 20 m/s.

(b) Calculate the kinetic energy of a car which has a mass of 1000 kg and is moving at the rate of 25 m/s.

(c) What is the speed of a horse weighing 345 kg and having a kinetic energy of 1.725×10^4 J?

4. The formula that can be used to find the compound interest is given by

$$A = P\left(1 + \frac{R}{n \times 100}\right)^{n \times T}$$

where A is the principal plus interest, P means Principal value, R is the rate of interest, T is time in years and n is the number of times interest is compounded per year.

(a) Amy invests $3500 in a bank for 5 years at 5.48% per annum, where the interest is compounded quarterly. Find the value of her investment at the end of the 5 years.

(b) Paul borrows $10800 for 4 years at 7.2% per annum, where the interest is compounded monthly. Find the interest paid.

5. Impulse (I) is defined as the change in the momentum of a body caused over a very short time. If m is the mass and v and u the final and initial velocities of a body, then
I = $m(v - u)$.

   (a) Calculate the impulse of a body of mass 5 kg whose speed increases from 10 m/s to 15 m/s in a short amount of time.

   (b) Calculate the mass of a body having an impulse of 480 Ns when its speed increases from 18 m/s to 30 m/s.

6. Elastic potential energy (PE) is defined as the Potential energy stored as a result of deformation of an elastic object. It is equal to the work done to stretch the spring, which depends upon the spring constant k as well as the distance stretched (L).
The formula used to calculate PE is given by
$PE = \frac{1}{2}kL^2$.

   (a) Determine how much Elastic Potential energy a spring with a spring constant of 12 N/m stores if it is stretched by 1.4m.

   (b) Determine the spring constant k for a spring having elastic Potential energy 27.04 J and stretched a distance of 2.6m.

MATHEMATICS APPLICATIONS UNIT 1

7. The period of a pendulum (T) can be worked out using the formula
$$T = 2\pi\sqrt{\frac{l}{g}}$$
where T is the time in seconds, l is the length of the pendulum (in metres) and g is the acceleration due to gravity in m/s² (use $g = 10$ m/s²)

(a) Use this formula to determine the period of a pendulum having length 0.60 m.

(b) What would be the period of a pendulum if it is 1.5m long?

(c) If the pendulum's length in (b) were to be shortened by one-third its original value, what would be its new period?

8. If m_1 and m_2 are the masses of two different objects and d is the distance between the centres of the two objects, then the gravitational attraction (F) between the two objects also known as Newton's Law of Gravity is given by $$F = \frac{Gm_1m_2}{d^2}$$
where G is the gravitational constant (Use $G = 6.7 \times 10^{-11}$).

(a) Two objects have masses 200 000 kg and 300 000 kg respectively. The centres of the two objects are 0.5m apart. Determine the gravitational attraction between the two objects.

(b) Explain what happens to the value of F as the distance between the objects decreases.

1E SPREADSHEETS

The objective of using spreadsheets is to learn how to use formulae to perform calculations and understand how we use cell addresses within a formula.

There are some important rules governing spreadsheets which we have to apply when solving problems.

- All formulae start with an = sign to identify them as a formula.
- We use the asterix (*) for multiplication
- For division make use of the forward slash (/)

EXAMPLE 1

We are going to use the formula $y = mx + c$ and a spreadsheet to calculate values of y for different sets of values of *m, x* and *c*.

Consider the second row, to calculate *y* enter **= A2*B2 + C2** in cell D2 as shown and the answer will appear as 14.

Similarly in cell D3 enter =A3*B3 + C3 and the value of 22 will automatically come up in that cell.

To fill the rest of the spreadsheet we can just drag and drop and the other cells D4, D5 etc.. will be worked out automatically.

	A	B	C	D
1	m	x	c	y
2	2	5	4	= A2*B2 + C2 14
3	3	4	10	= A3*B3 + C3 22
4	5	3	-7	**8**
5	10	-2	-5	**-25**
6	8	4	6	**38**

EXAMPLE 2

The spreadsheet below shows the T-Shirt sales of Top Clothing Ltd. Insert appropriate formula in cells G2, G3 and so on to find the total number of shirts sold for each size.

Consider the second row, to calculate the total enter **= (sum B2 : F2)** in cell D2 as shown and the answer 76 will be displayed. The rest of column G can as usual be dragged and filled.

		B	C	D	E	F	G
1		Monday	Tuesday	Wednesday	Thursday	Friday	Total
2	X Small	12	15	10	21	18	=(sum B2 : F2) 76
3	Small	10	11	8	13	12	54
4	Medium	25	21	33	41	38	158
5	Large	8	7	5	10	6	36

EXERCISE 1E

1. Use a spreadsheet from your calculator to complete the following.

	A	B	C	D
1	Quantity sold	Price	Sales	GST
2	8	5	= A2*B2 40	= 0.1*C2 4
3	10	9		
4	15	4		
5	7	20		
6	22	3		

2. The spreadsheet below shows the marks of a group of students studying Application Mathematics in four different tests. The tests were marked out of 10 marks each. Use appropriate formula in your spreadsheet to complete columns F and G.

		B	C	D	E	F	G
1		Test 1	Test 2	Test 3	Test 4	Total Marks	%
2	Alice	9	7	5	8		
3	Berry	5	4	6	4		
4	Carol	9	9	8	9		
5	Damien	10	8	8	5		
6	Essen	2	4	5	1		

3. The area of a trapezium is $A = \frac{a+b}{2} \times h$. Use the spreadsheet below and your calculator to complete the table for different sets of values of *a*, *b* and *h*.

	A	B	C	D
1	*a*	*b*	*h*	Area
2	8	6	10	
3	11	19	5	
4	15	17	8	
5	7	23	20	
6	19	21	8	

CHAPTER 2

PERCENTAGES

A percentage is indicated by the symbol %, and it means a fraction of one hundred.

Example $\quad 13\% = \frac{13}{100} = 0.13$

It is customary to express a fraction in its simplest form.

Example $\quad 15\% = \frac{15}{100} = \frac{3}{20}$

CLASS ACTIVITY

Complete the following table

Percentage	Fraction	Decimal	Percentage	Fraction	Decimal
17%	$\frac{17}{100}$	0.17	60%	$\frac{60}{100} = \frac{3}{5}$	0.6
30%				$\frac{12}{25}$	
	$\frac{31}{100}$				0.65
		0.4	64%		
36%				$\frac{13}{50}$	
210%	$2\frac{10}{100} = 2\frac{1}{10}$	2.1	325%		
		1.45		$4\frac{1}{5}$	

2A   EXPRESSING ONE QUANTITY AS A PERCENTAGE OF ANOTHER

A percentage can be used to compare two quantities expressed in the same units.
Consider a Year 11 class of 24 students studying Mathematics Applications.
If 16 students live in the same suburb, we can express the students living in the same suburb as a percentage of the whole class.

Thus fraction of students living in the same suburb $= \dfrac{\textit{Number of students living in the same suburb}}{\textit{total number of students in the class}}$.

To convert fraction into percentage, we multiply by 100 as usual.

Percentage of students living in the same suburb $= \dfrac{16}{24} \times 100 = 66.7\ \%$

EXAMPLE

1. What percentage is 32 out of 50? **SOLUTION** $$\dfrac{32}{50} \times 100 = 64\ \%$$	2. Express 400 metres as a percentage of 2.45 km. **SOLUTION** 2.45 km = 2450 m $$\dfrac{400}{2450} \times 100 = 16.33\ \%$$
3. Express 93g as a percentage of 3 kg. **SOLUTION** 3 kg = 3000 g $$\dfrac{93}{3000} \times 100 = 3.1\ \%$$	4. Express 2500 m² as a percentage of 12 hectares (1 ha = 10 000m²). **SOLUTION** 12 ha = 120000 m² $$\dfrac{2500}{120000} \times 100 = 2.1\ \%$$
5. An airline company has seats for 120 passengers. Calculate (a) The number of passengers on board when $\dfrac{7}{15}$ of the seats are occupied. $$\dfrac{7}{15} \times 120 = 56$$ (b) The percentage of seats which are occupied when there are 114 passengers on board. $$\dfrac{114}{120} \times 100 = 95\%$$	

EXERCISE 2A

1. What percentage is 11 out of 20?	2. Express 400 metres as a percentage of 3.6 km.
3. Express 120g as a percentage of 5 kg.	4. Express 36000 m^2 as a percentage of 15 hectares.
5. A theatre has seats for 1100 people. Calculate the percentage of seats that are occupied if 847 people attend.	6. To make a chicken and mushroom pie, Mrs Mac requires 550g of chicken and 125g of mushroom. Express the mass of mushroom as a percentage of the mass of the chicken, giving your answer correct to the nearest whole number.
7. A market gardener sows 25000 lettuce seeds. Given that 21500 produce seedlings, calculate the percentage which did **not** produce seedlings.	8. In September 2012 the number of students at a school was 540. Given that the number of students at the school in September 2013 was 567, find the percentage increase in the number of students.
9. Jessy earns $1500 each month. He pays rent of $525 each month. Find the amount he pays in rent as a percentage of his earnings.	10. In a sale, the price of bicycle B is reduced from $2400 to $1596. Calculate the percentage reduction given.

2B FINDING A GIVEN PERCENTAGE OF AN AMOUNT

To find a given percentage of an amount, use the following steps:

- Write the given percentage as a fraction
- Multiply the amount by the fraction

EXAMPLE

1. Find 12% of 300. **Solution** $12\% \text{ of } 300 = \dfrac{12}{100} \times 300 = 36$	2. Find 15% of $10.80 **Solution** $15\% \text{ of } \$10.8 = \dfrac{15}{100} \times 10.80 = \$1.62$

EXERCISE 2B

1. Find 13% of 400.	2. Find 8% of $25.60
3. Find 125% of 200 kg.	4. Find 24% of 90 litres.
5. Find 21% of $800.	6. Find 20% of $500
7. On a packet of tomato seeds it is stated that 93% should produce seedlings. Find the number of seedlings which would be expected from a packet containing 600 seeds.	8. Of a class of 60 students, 25% play soccer, 35% play tennis and the rest play netball. Calculate the number of students who play netball.

2C INCREASE AND DECREASE IN PERCENTAGES

To increase or decrease an amount by a certain percentage, it is easier to multiply the amount by a multiplying factor as shown below.

CLASS ACTIVITY

Complete the table.

Increase by	Multiplying factor	Decrease by	Multiplying factor
8%	100% + 8% = 108% = 1.08	12%	100% − 12% = 88% = 0.88
15%	1.15	2.5%	0.975%
7%		4%	
13%		15%	
125%		90%	
4.5%		4.5%	
8.5%		0.5%	

EXAMPLE

1. Increase 80 by 10%.
   SOLUTION
   $80 \times 1.10 = 88$
	2. Decrease 200 by 5%.
	   SOLUTION
	   $200 \times 0.95 = 190$

3. Due to an increase in the cost of living, Tom's salary increased from $16.80 per hour by 3%. Calculate Tom's new hourly rate.
   SOLUTION
   $16.80 \times 1.03 = \$17.30$
	4. In 2013 the number of tourists visiting Monkey Mia was 15600. This number decreases by 2.5 % in 2014. Find the number of tourists who visited Monkey Mia in 2014.
	   SOLUTION
	   $100 - 2.5 = 97.5\% = 0.975$
	
	   $15600 \times 0.975 = 15210$

EXERCISE 2C

1. Increase 300 by 11%.	2. Decrease $400 by 6%.
3. Increase 500 kg by 15%.	4. Decrease 4600 m by 7%.
5. Increase 1600 km by 9%.	6. Decrease $1600 by 3.5%.
7. Increase 55 kg by 2.3%.	8. Decrease 450m by 23%.
9. A clerk's monthly salary is $5600. If his monthly salary increases by 4.2%, calculate his new monthly salary.	10. In 2014, the number of tourists visiting Perth increased by 5% compared to 124000 in 2013. Calculate the number of tourists visiting Perth in 2014.
11. A factory produces 15000 toys each year. Due to a slowdown in the economy, the manager decides to reduce the production capacity by 7.5%. Calculate the new production level.	12. A man earning $5625 is awarded a pay rise of 8%. Calculate his new salary wage.
13. The price of X-Box 360 decreases from $360 to $280. What is the percentage decrease in price?	14. In 2013, the cost of posting a letter was increased from 60 cents to 72 cents. Calculate the percentage increase.

2D CALCULATING AN ORIGINAL AMOUNT

EXAMPLES

1. 15% of an amount is 40, find the amount. **SOLUTION** Amount = $40 \times \frac{100}{15} = 266.67$	2. 5% of a ticket cost $2.40. Find the cost of the ticket. **SOLUTION** Cost of ticket = $2.40 \times \frac{100}{5} = 48$
3. Hotel expenses, totalling $234, accounts for 65% of the cost of the holiday. Calculate the total cost of the holiday. **SOLUTION** Cost of holiday = $234 \times \frac{100}{65} = \$360$	4. There were 756 children in a school. This is 5% more than it was last year. Calculate the number of children in the school last year. **SOLUTION** Number of children = $756 \times \frac{100}{105} = 720$

EXERCISE 2.5

1. 12% of an amount is 66, find the amount.	2. 8% of a cinema ticket cost $1.92. Find the cost of the ticket.
3. 35% of an amount is 210, find the amount.	4. 25% of a number is 72. Find the number.
5. If 121 students went on holiday last year, accounting for 11% of a school population, find the number of students at the school.	6. There were 903 children in a school. This is 5% more than it was last year. Calculate the number of children in the school last year.
7. Tax on the original price of a bicycle is charged at 20% of the original price. After tax has been included, Matthew pays $1080 for this bicycle. Calculate the original price.	8. In June 2012, there were 450 members at a tennis club. Given that this was 20% more than that in June 2011, find the number of members in June 2011.

2E PROFIT AND LOSS

A businessman makes a profit on selling an article if the selling price is more than the cost price. On the other hand, a businessman who sells an article at a price lesser than the cost price incurs a loss.

Hence   Profit = Selling Price – Cost Price
         Loss = Cost Price – Selling Price

EXAMPLES

1. A trader bought a calculator for $60 and sold it for $75. Calculate the percentage profit.
   SOLUTION
   Profit = $75 - $60 = $15
   Percentage profit = $\frac{profit}{cost\ price} \times 100$
   $= \frac{15}{60} \times 100 = 25\%$

2. Alex bought a DVD player for $90 and sold it for $80. Calculate his percentage loss.
   SOLUTION
   Loss = $90 – $80 = $10
   Percentage loss = $\frac{loss}{cost\ price} \times 100$
   $= \frac{10}{90} \times 100 = 11.1\%$

3. For her formal, Ashley bought a dress on EBay for $400 and sold it a year later at a profit of 8%. Calculate the selling price of the dress.
   SOLUTION
   Selling price = $1.08 \times 400 = \$432$

4. JR Trading sold a toy for $76.50 incurring a loss of 15%. Find the cost price of the toy.
   SOLUTION
   Cost price = $76.50 \times \frac{100}{85} = \$90$

5. During the end of year financial sale, all prices are reduced by 40%. Calculate the original price of an article whose sale price is $48.
   SOLUTION
   Original price = $48 \times \frac{100}{60} = \$80$

6. When Tony, the salesman, sells a camera for $84 he makes a profit of 12%. Calculate the cost price of the camera.
   SOLUTION
   Cost price = $84 \times \frac{100}{112} = \$75$

7. Matthew makes pieces of furniture and sends them to a shop where they are sold.
   When a piece is sold, the shopkeeper receives 15% of the selling price, and Matthew receives   the rest.

(a) A table is sold for $200.
(i) Calculate the amount the shopkeeper receives.
   $\frac{15}{100} \times 200 = \$30$
(ii) The cost of making this table was $131.80.
Calculate the percentage profit that Matthew makes when this table is sold.

Profit = $200 – $131.80 – $30 = $38.20
Percentage profit = $\frac{38.20}{131.80} \times 100 = 28.98\%$

(b) Matthew made a bookcase.
The cost of making the bookcase was $647.50. After the bookcase is sold and the shopkeeper has received 15% of the selling price, Matthew makes a profit of $160.
Calculate the selling price of the bookcase.

Selling price = $(647.50 + 160) \times \frac{100}{85} = \$950$

CHAPTER 2 : PERCENTAGES

CLASS ACTIVITY
Complete the table below.

Cost Price	Selling Price	Profit/Loss	Percentage profit/ Percentage Loss
$20	$24	Profit = $4	% profit = $\frac{profit}{cost\ price} \times 100$ $= \frac{4}{20} \times 100 = 20\%$
$200	$190		
$460	$598		
$50	$23		

Find the selling price in each of the following cases.

1. Cost price $60, profit 15%	2. Cost price $500 , loss 8%
3. Cost price $1200, profit 2.5%	4. Cost price $5400, loss 3%
5. Cost price $2000, profit 5%	6. Cost price $72, loss 4%

Find the cost price in each of the following cases.

7. Selling price $550, profit 10%	8. Selling price $558, profit 24%
9. Selling price $680, loss 10%	10. Selling price $1200, profit 20%

EXERCISE 2E

1. A man bought a picture for $325 and sold it at a profit of 12%. Calculate the selling price.

2. A dealer made a profit of 20% by selling a car for $6300. Calculate the price he paid for the car.

3. The owner of a toy shop made a profit of 40% on every toy which he sold.

   (a) Find the selling price of a soccer ball which cost $11.

   (b) The cost price of a Barbie doll which was sold for $28.

   (c) The shopkeeper made a profit of $10 when he sold a kite. Calculate the selling price of the kite.

4. Jordan usually sells refrigerators for $96 each. He discovers that one refrigerator has been slightly damaged, so he reduces the price by 15%.

   (a) Calculate the new selling price for the damaged refrigerator.

   (b) When he sells the refrigerator at the usual price of $96, Jordan makes a profit of 20%. Calculate the cost price of the refrigerator.

5. In a sale, a shop reduces all its prices by 20%.

   (a) Calculate the cost of an article which was originally priced at $26.

   (b) The original price of an article which was sold in the sale for $44.

6. (a) A bookseller bought a gardening book for $7.50 and sold it for $10.50. Calculate the percentage profit made by the bookseller.

   (b) During a sale the bookseller reduced the price of his books by 30%. Calculate the normal selling price of a book which was priced in the sale at $13.30.

2F   DISCOUNT

Discount is a reduction from the usual cost of a product. For example, some customers get a discount for buying in bulk. Some businesses give discount to attract customers to buy more and thus increasing their sales.

EXAMPLES

1. A new television set cost $800. For the month of June a discount of 5% is allowed to all buyers. Calculate the cost of the television in June. **SOLUTION** $800 \times 0.95 = \$760$	2. Joe paid $108 for a set of books after a discount of 10%. Calculate the original price of the set of books. **SOLUTION** $108 \times \dfrac{100}{90} = \$120$

EXERCISE 2F

Find the selling price of each of the following after the discount.

1. Original price $200, 15% discount	2. Original price $300, 12% discount
3. Original price $600, 5% discount	4. Original price $2000, 2.5% discount
5. Original price $450, 24% discount	6. Original price $8200, 4% discount

Find the original price in each of the following.

7. A sum of $132 is paid after a discount of 12%.	8. A sum of $3551.20 is paid after a discount of 8%.
9. A sum of $23040 is paid after a discount of 4%.	10. A sum of $2375 is paid after a discount of 5%.

11. A shop is offering discounts on shirts costing $30 each. If someone buys 2 shirts, he will be offered a discount of 12% on the first shirt and another 10% discount on the reduced price for the second shirt. How much would one pay for two shirts at this shop?

12. The price of a pair of Nike socks was decreased by 22% to $30. What was the original price of the socks?

13. (a) In 2013 the cost of posting a letter was 36 cents.
(i) A company posted 3000 letters and was given a discount of 4%.
Calculate the total discount given. Give your answer in dollars.

(ii) In 2014, the cost of posting a letter was increased from 36 cents to 45 cents.
Calculate the percentage increase.

(iii) After the price increase to 45 cents, the cost to the company of posting 3000 letters was $1302.75. Calculate the new percentage discount given.

(b) In 2014, it cost $5.40 to post a parcel. This was an increase of 12.5 % on the cost of posting the parcel in 2014. Calculate the increase in the cost of posting this type of parcel in 2014 compared to 2013.

2G COMMISSION

Commission is money that a person earns based on how much he or she sells. It is usually expressed as a percentage of the total amount sold.

The following people normally work on commission as well as a retainer (fixed wage) : Real estate agents, car dealers, and pharmaceutical sales representatives . For example, a real estate agent earns a fraction of the selling price of a house that he or she helps a customer to purchase or sell. Similarly, a car dealer earns a portion of the selling price of an automobile that he or she is able to sell.

EXAMPLE 1

Patrick, an art dealer, earns 12% commission of the dollar value of the art pieces that he sells at the Louvre. Patrick sold $42000 worth of art pieces this month. How much commission does he receive?

SOLUTION

Commission = $0.12 \times 42000 = \$5040$

EXAMPLE 2

Alex is a car dealer and he earns 15% commission of his exclusive range of luxury vehicles sales. Last year, his salary was $48000. What was the total dollar amount of his sales last year?

SOLUTION

Total sales = $48000 \times \frac{100}{15} = \$320000$.

EXAMPLE 3

George is employed by a real-estate agency and is paid commission as follows:

First $ 200 000 → 4 %

Amount exceeding $ 200 000 → 2.5%

George sold a property worth $ 280 000. Calculate the total commission paid to him.

SOLUTION

First $ 200 000 → $0.04 \times 200\,000 = \$8000$

Amount exceeding $ 200 000 (280 000 − 200 000 = 80 000) → 2.5% × 80 000 = $ 2000

Total commission = $ 8000 + $ 2000 = $ 10 000

MATHEMATICS APPLICATIONS UNIT 1

EXERCISE 2G

1. Chloe sells $2500 worth of toys, and she makes 20% commission. How much money did she make?

2. Dane's weekly pay is $480, and he gets 10% commission on his sale. If he sells $2800 worth of goods, how much money does he get in total?

3. A salesman earns $3000 per month plus 2% commission on his sales. If he sells goods worth $24000 in one month, how much does he earn in that month?

4. Jenna is a salesperson at an electronics store. She earns 6.5 % commission on her total sales. What would be her commission if she sold a $2950 plasma television set?

5. Clara makes $800 per month plus 8% commission. What must commission sales reach for her total salary to be $2000?

6. Alexandra makes $7 an hour plus 12% commission selling jewellery. How much does she earn on an 8 hour shift in which she sells $450 worth of jewellery?

7. John, a computer salesperson earns $35 000 as a retainer each year. This year John sold $240 000 worth of computers. If a commission of 2.5% is paid on the sales, find John's income for the year.

8. Peter receives a flat salary of $3000 a month plus a commission of 2% for the value of goods he sells. During a particular month he received $3360 in total. Calculate the total value of goods he sold that month.

9. Jean Luc, an art dealer, earns 15% commission of the dollar value of the art pieces that he sells at the Australian Academy of Arts. Jean Luc earns $9600 this month. What is the total dollar value of the art that he sells?

10. Roger is an agent for movie stars. He earns 12% of his clients' salaries. If he made $75,000 last year, how much did his clients make in all?

11. Matilda works as a salesgirl in a jewellery store on Murray Street. She is paid a 9.5% commission on her sales. One very busy day she made the following 3 sales;
A ladies watch: $275.95
A diamond necklace: $599.99
A pair of cufflinks: $125
What was Matilda's commission on her total sales?

12. George is employed by a real-estate agency and is paid commission as follows :

First $ 200 000 → 2.5 %
Next $ 100 000 → 2 %
Amount exceeding $ 300 000 → 1.5%

George sold a property worth $ 360 000.
Calculate the amount of commission he earned.

13. A salesperson receives step commission on sales calculated as follows:

* 8% on first $1000
* 10% next $2000
* 15% on sales above $3000

Calculate the salesperson's earning if in one week his sales was $8500.

2H CALCULATING TAX PAID AND TAXABLE INCOME

EXAMPLE

The current income tax rates for individuals in Australia are shown in the table below.

Taxable income	Tax on this income
0 - $18 200	Nil
$18 201 - $37 000	19c for each $1 over $18 200
$37 001 - $80 000	$3 572 plus 32.5c for each $1 over $37 000
$80 001 - $180 000	$17 547 plus 37c for each $1 over $80 000
$180 001 and over	$54 547 plus 45c for each $1 over $180000

Use the table to determine the income tax paid by an individual who has a taxable income of

(a) $15500

$$tax\ paid = \$0$$

(b) $27500

$$Tax\ paid = 0.19(27500 - 18200) = \$1767$$

(c) $85400.

$$Tax\ paid = 17547 + 0.37(85400 - 80000) = \$19545$$

(d) How much does the individual in part (c) take home monthly?

$$\frac{85400 - 19545}{12} = \$5487.92$$

(e) Julian paid $15064 in tax. Determine Julian's taxable income.

Clearly Julian must in the $37 001 - $80000 tax brackets.

Solve $(3572 + 0.325(x - 37000) = 15064)$ on CAS

$$x = \$72360$$

Therefore Julian's taxable income is $72360.

EXERCISE 2H

Income tax rates for 2014–15 for both Australian citizens and foreign residents working in Australia are given in the tables below. The rates apply from 1 July 2014.

Taxable income	Tax on this income
0 - $18 200	Nil
$18 201 - $37 000	19c for each $1 over $18 200
$37 001 - $80 000	$3 572 plus 32.5c for each $1 over $37 000
$80 001 - $180 000	$17 547 plus 37c for each $1 over $80 000
$180 001 and over	$54 547 plus 45c for each $1 over $180 000

1. Use the table to determine the income tax paid by Alex who has a taxable income of $54500.

2. James has a taxable income of $105 000. Determine the income tax he paid.

3. Use the table to determine the income tax paid by Charles who has a taxable income of $32000.

4. Dave has a taxable income of $190000. Use the table to determine the income tax he pays.

5. Calculate the taxable income of Eric who paid $23615 in tax in the 2014–15 financial year.

6. Calculate the taxable income of Fredric who paid $61927 in tax in the 2014–15 financial year.

7. From 1 July 2013, these fees were recommended by the Roswell Real Estate Company.

Real Estate Fees/Commission in Roswell

Selling price of each property	Real estate fees
Does not exceed $10 000	10.2% with a minimum of $100
From $10 001 – $50 000	$1050 plus 5.8% of excess over $10 000
From $50 001 – $100 000	$3400 plus 4.1% of excess over $50 000
From $100 001 upward	$5460 plus 3.9% of excess over $100 000

Alexa owns several properties in Roswell.

(a) If Alexa sells a house for $280 000, how much does she have to pay in real estate fees?

(b) In a particular week Alexa sells two properties, a house for $198 000 and a unit for $99 900. Calculate the total fees she needs to pay.

(c) Alexa wants to reduce the number of properties she owns. She is trying to decide whether to sell a house for $280 000 or sell a group of four apartments, for $70 000 each. Which option will result in her paying the smaller amount of real estate fees and how much less will she pay in fees with this option?

CHAPTER 3

SIMPLE INTEREST

3A CALCULATING SIMPLE INTEREST

To find the simple interest (I), use the formula I = P R T, where P is the principal value or present value, R is the rate of interest and T is time in years.

EXAMPLE 1

Find the simple interest to be earned on each of these investments.

(a) $4000 for 3 years at 9% p.a.

$4000 \times \frac{9}{100} \times 3 = \$ 1080$

(b) $2000 for 9 months at 6.5% p.a.

$2000 \times \frac{6.5}{100} \times 0.75 = \$ 97.50$ [9 months is ¾ of a year or simply divide 9 by 12]

EXAMPLE 2

A family borrows $16000 to buy a car and are charged 15% p.a. simple interest. If the loan is for 5 years find:

(a) the amount of interest to be paid

PRT = 16000 × 0.15 × 5 = $ 12000

(b) the total amount to be repaid

$16000 + $ 12000 = $ 28000

(c) the amount of each repayment if the loan is to be repaid in 60 equal instalments.

28000 ÷ 60 = $ 466.67

EXAMPLE 3

Peter intends to buy a new car costing $24000 for his 21st birthday. He saved 30% of the amount needed in the bank. He decided to borrow the remainder from MeBank and agree to make quarterly repayments, with simple interest to be paid at the rate of 12% per annum. Including the principal, Peter will pay the debt in full in 5 years. Calculate Peter's quarterly repayments.

SOLUTION

Amount borrowed = 70% of $ 24000 = $ 16800

Interest payable = 16800 × 0.12 × 5 = $10080

Total amount to be repaid = 16800 + 10080 = 26880

5 years = 5 × 4 = 20 quarters ∴ Quarterly payments = 26880 ÷ 20 = $ 1344

EXERCISE 3A

Find the simple interest to be earned on each of these investments.

1. $500 for 2 years at 16% p.a.	2. $2 500 for 4 years at 1.2% p.a.
3. $16 450 for 3 years at 8% p.a.	4. $350 for 210 days at 42% p.a.

5. How much interest is payable if I borrow $6 000 at 7.4% for 21 months?

6. $15 000 is invested at 6% p.a. simple interest for 10 years.
   a   How much interest will be earned each year?

   b   How much interest will be earned over the 10-year period?

7. A family borrows $850 to buy a television and are charged 18% p.a. simple interest. If the loan is for 2 years find:
   a   the amount of interest to be paid

   b   the total amount to be repaid

   c   the amount of each repayment if the loan is to be repaid in 24 equal instalments.

8.

Account	Simple interest per year
Alpha saver	4.2%
Beta saver	4.4%

On 31 December 2011, Kelly and Johan each had $6000 in an account.

Kelly's money is in a Alpha Saver Account.
Johan's money is in a Beta Saver Account.

(i) How much money did Kelly have in her account on 31 December 2012 after the interest had been added?

(ii) On 31 December 2012, Kelly transferred this money to the Beta Saver Account. How much money did she have in this account on 31 December 2013 after the interest had been added?

(iii) Johan kept her money for the two years in the Beta Saver Account, which earned simple interest of 4.4% per year.
After all interest had been added, who had more money in their account on 31 December 2013 and by how much?

3B   SIMPLE INTEREST AND TECHNOLOGY (CAS)

Simple interest can also be calculated by using the class pad calculator.

- Menu → Financial → Simple interest

EXAMPLE 1

Use the class pad to find the interest obtained when $6000 is invested for 3 years at 8% per annum.

SOLUTION

Days	365 × 3 = 1095
I%	8
PV	6000
SI	Solve and 1440 appears (**ignore the minus sign**)
SFV	If you solve this gives 6000 + 1440 = 7440 which is the interest added to the principal value

EXERCISE 3B

Using your calculator, find the simple interest to be earned on each of these investments.

1. $2500 for 4 years at 12% p.a.	2. $4 500 for 2 years at 6.5% p.a.
3. $16 300 for 120 days at 8% p.a.	4. $45000 for 5 years at 8.5% p.a.
5. Calculate the simple interest when $ 5400 is invested for 6.5 years at 8.8% per annum.	6. Calculate the simple interest when $6500 is invested for 18 months at 7.6% per annum.

3C FINDING PRINCIPAL (P), RATE OF INTEREST (R) OR TIME (T)

If three of the four variables in the formula I = P R T are known, we can calculate the fourth one by simply rearranging the formula or using CAS as show in the examples below.

$$P = \frac{100\,I}{R \times T}, \quad R = \frac{100\,I}{P \times T} \quad \text{and} \quad T = \frac{100\,I}{P \times R}$$

EXAMPLE

Joseph invested $3000 for 4 years and his savings accumulated to $3600. Calculate the rate of interest.

SOLUTION **Alternative 1** Using $R = \frac{100\,I}{P \times T} = \frac{100 \times 600}{3000 \times 4} = 5$ Rate of interest = 5%	**Alternative 2 (CAS)** Main→ Action →Advanced → Solve Using I = P R T Solve (600 = 3000 × x × 4) x = 0.05 rate of interest = 5%

EXERCISE 3C

Find the principal invested in each of *these* simple interest investments.

1. Interest of $400 at 8% for 5 years.	2. Interest of $6200 at 3.2% for 2 years
3. Interest of $5 000 at 7% p.a. for 2 years	4. Interest of $1 400 at 9.2% for 3 years
5. An investor has $50 000 invested in an account that pays 4.6% p.a. simple interest. If she wants to earn at least $20 000 in interest, for how many years will the money need to be invested?	6. Darren has received $90 000 in interest payments on an investment of $500 000 that he made 4 years ago which paid simple interest. What rate of interest has been paid?

7. A sum of $350 is invested at simple interest and amounts to $476 after 3 years. Calculate
   (a) The total interest earned,

   (b) The rate percent per annum.

8. Calculate the simple interest when $2800 is invested for 9 months at 8% per annum.

9. Lily, Margaret and Nancy were each left $8000 in their aunt's will. Margaret invested her money in a bank at 9.5% simple interest for 5 years. Calculate the total amount of money she had in the bank after 5 years.

10. Find the simple interest obtained when $12500 is invested at 7.2% for a period of 27 months.

11. A bank charges $28 simple interest on a sum of money which is borrowed for four months. Given that the rate of interest is 15% per annum, calculate the sum of money.

12. How long does it take $12000 to yield a simple interest of $2640 at 5.5% per annum?

13. What sum of money will yield a simple interest of $5400 at 9% per annum in 3 years?

14. Joseph has a deposit of $16 000 in an investment account which pays a simple interest rate of 5% p.a. How long would it take him to earn $5 200 in interest?

CHAPTER 4

COMPOUND INTEREST

4A CALCULATING COMPOUND INTEREST

Compound interest is different from Simple Interest as it involves adding interest at regular intervals to the amount invested or borrowed.

The formula that can be used to find the compound interest is given by

$$A = P\left(1 + \frac{R}{n \times 100}\right)^{n \times T}$$

- A is the principal plus interest
- P means Principal value,
- R is the rate of interest,
- T is time in years and
- n is the number of times interest is compounded per year. For example, if interest is compounded quarterly n = 4.

EXAMPLES

1. Emily invests $2450 in a bank for 4 years at 4.8% per annum, where the interest is compounded. Find the **value of her investment**, if the interest is paid:

   (a) annually

   $A = 2450\left(1 + \frac{4.8}{1 \times 100}\right)^{1 \times 4}$

   $= \$2955.37$

   (b) monthly

   $A = 2450\left(1 + \frac{4.8}{12 \times 100}\right)^{12 \times 4}$

   $= \$2967.46$

   (c) every six months

   $A = 2450\left(1 + \frac{4.8}{2 \times 100}\right)^{2 \times 4}$

   $= 2961.87$

2. Robredo invests $10000 in a bank for 6 years at 8% per annum, where the interest is compounded. Find the *interest earned* on his investment, if the interest is paid:

   (a) quarterly

   $A = 10000\left(1 + \frac{8}{4 \times 100}\right)^{4 \times 6}$

   $= \$16084.37$

   Interest = $16084.37 - 10000$

   $= \$6084.37$

   (b) daily

   $A = 10000\left(1 + \frac{8}{365 \times 100}\right)^{365 \times 6}$

   $= \$16159.89$

   Interest = $16159.89 - 10000$

   $= \$6159.89$

EXERCISE 4A

1. Alex invests $9600 in a bank for 5 years at 9% per annum, where the interest is compounded. Find the value of his investment, if the interest is paid:

(a)	Annually	(b)	each 6 months
(c)	each month	(d)	quarterly

2. **Susan** invests $12000 in a bank for 3 years at 10% per annum, where the interest is compounded. Find the *interest earned* on her investment, if the interest is paid:

(a)	Annually	(b)	each 6 months
(c)	each month	(d)	quarterly

4B COMPOUND INTEREST AND TECHNOLOGY

Compound interest can also be calculated by using the class pad. Use the following steps:

- Menu
- Financial
- Compound interest

EXAMPLE 1		EXAMPLE 2	
Peter invests $6000 in a bank for 5 years at 9% per annum, where the interest is compounded. Find the value of his investment and the interest earned, if the interest is paid monthly. **SOLUTION** To obtain the meaning of each of the letters, click help		Ashley invests $5600 in a bank for 3 years at 8.4% per annum, where the interest is compounded. Find the value of his investment and the interest earned, if the interest is paid quarterly. **SOLUTION**	
N	12×5 = 60	N	4 × 3 = 12
I%	9	I%	8.4
PV	6000	PV	5600
PMT	0	PMT	0
FV	SOLVE and we get 9394.09 (The answer would appear negative, ignore it)	FV	SOLVE and we get 7186.16 (The answer would appear negative, ignore it)
P/Y	12	P/Y	4
C/Y	12	C/Y	4
The value of his investment is $9394.09 and the interest earned = $9394.09 – 6000 = $3394.09		The value of her investment is $7186.16 and the interest earned = $7186.16 – 5600 = $1586.16	

EXERCISE 4B

Use your class pad calculator (**financial**) to answer the following questions

1.	John invests $8500 in a bank for 6 years at 12% per annum, where the interest is compounded. Find the value of his investment, if the interest is paid:	(a)	annually
		(b)	quarterly
		(c)	monthly
2.	Peter borrows $12500 for 4 years at 9.6% per annum, where the interest is compounded. Find the interest paid if the interest is paid:	(a)	annually
		(b)	quarterly
3.	Alex invests $64000 in a bank for 8 years at 6.5% per annum, where the interest is compounded. Find the value of his investment, if the interest is paid:	(a)	quarterly
		(b)	monthly

4. John wishes to invest $60 000 for 5 years. Three options are being considered.

    Option A: Simple interest at 8.6 % per year

    Option B: Compound interest at 7% per annum compounded yearly

    Option C: Compound interest at 7.5% per annum compounded quarterly

Which investment should John choose to maximise the interest earned?

CHAPTER 4 : COMPOUND INTEREST

5. Sarah invests $1 250 in a bank for 3 years at 4.8% per annum, where the interest is compounded. Find the value of her investment, if the interest is paid quarterly.

6. The West Bank offers an account in which compound interest is calculated every quarter. The interest rate is 5.2% per annum. The South Bank pays 5.4% compound interest and is calculated yearly. Philip has inherited $7000 from his grandmother's will and plans to invest the money for a period of 5 years. Which bank will offer him more interest?

7. Casey has $115 000 and wishes to invest it for 4 years. She checks out two banks and is offered the following terms:

   Bank A: 5.5% p.a. compounded yearly

   Bank B: 5.3 % p.a. compounded monthly.

   Determine which bank offers the best deal.

4C HOW MUCH DO I REPAY?

We can also use the inbuilt financial programmes to determine the amount to be repaid by an individual when the latter borrows from any financial institution. The steps are same as before.

- Menu
- Financial
- Compound interest

EXAMPLE

Cyrus borrowed $20000 with Delta Bank with compound interest of 14% per annum, the amount to be paid off in 5 years. Calculate the constant amount to be repaid (PMT) each month for Cyrus to see his loan to be written off.

SOLUTION

N	12×5 = 60
I%	14
PV	20000
PMT	Solve to obtain - $465.37
FV	0
P/Y	12
C/Y	12

Cyrus has to make a monthly repayment of $465.37.

EXERCISE 4C

1. Amanda borrowed $15000 with Alpha Bank with compound interest of 16% per annum, the amount to be paid off in 4 years. Calculate the constant amount to be repaid each month for Amanda to see her loan written off.	2. Calculate to the nearest cent the constant monthly repayment if $240 000 is loaned for 20 years with compound interest of 6.1% per year.
3. Calculate to the nearest cent the constant weekly repayment if $60000 is loaned for 10 years with compound interest of 9.5% per year.	4. Calculate to the nearest cent the constant yearly repayment if $8000 is loaned for 6 years with compound interest of 18% per year.

4D INFLATION

Inflation can be defined as the rate at which the general level of prices for goods and services is rising, and, subsequently, purchasing power is falling. Central banks in most countries attempt to minimise severe **inflation**, along with severe deflation, in an attempt to keep the excessive growth of prices to a minimum.

CLASS ACTIVITY : Copy and complete the table below.

Rate of inflation	Workings	Used in answering questions
5%	100 + 5 = 105%	1.05
7%		
13%		
20%		
4.8%	100 + 4.8 = 104.8 %	1.048
8.8%		
5.25%		

EXAMPLE 1

A gold necklace is bought for $3000 and gains 6% of its value each year. What is the value after:

(i) 3 years   (ii) 10 years   (iii) n years

SOLUTION
(i) $3000 (1.06)^3 = \$ 3573.05$
(ii) $3000 (1.06)^{10} = \$5372.54$
(iii) $3000 (1.06)^n$

EXAMPLE 2

An item is bought for $5000 and is valued at $6800 after 5 years. Find the constant rate of inflation that would allow this rise in value.

SOLUTION

$5000 x^5 = 6800$ (Main, Action, Advanced, Solve)

$x = 1.063$

Inflation rate = 6.3%

EXERCISE 4D

1. An item is bought for $4000 and gains 5% of its value each year. What is the value after: **(i)** 4 years **(ii)** 10 years **(iii)** n years	**2.** A house is bought for $250 000 and gains 4.5% of its value each year. What is the value after: **(i)** 5 years **(ii)** 15 years **(iii)** n years
3. A piece of jewellery is bought for $1200 and gains 6.25% of its value each year. What is the value after 3 years?	**4.** An item is bought for $12000 and is valued at $14000 after 3 years. Find the constant rate of inflation that would allow this rise in value.
5. An item is bought for $10000 and is valued at $12500 after 5 years. Find the constant rate of inflation.	**6.** An item is bought for $400 and doubles its value after 6 years. Find the constant rate of inflation.

4E   DEPRECIATION

Depreciation is the decline in the value of an item due to wear and tear. For example, the value of a car normally depreciates overtime.

Copy and complete the table below.

Rate of depreciation	Workings	Used in answering questions
5%	100 - 5 = 95%	0.95
2%		
12%		
25%		
4.5%	100 – 4.5 = 95.5%	0.955
8.5%		
5.65%		

EXAMPLE 1

A car is bought for $24 000 and loses 9% of its value each year. What is the value after:

(i) 3 years   (ii) 10 years   (iii) n years

SOLUTION

(i) $24000 (0.91)^3 = \$18085.70$
(ii) $24000 (0.91)^{10} = \$9345.99$
(iii) $24000 (0.91)^n$

EXAMPLE 2

An item is bought for $5000 and is valued at $3500 after 5 years. Find the constant rate of depreciation that would allow this drop in value.

SOLUTION

$5000 x^5 = 3500$ (Action, Advanced, Solve on your CAS calculator)

$x = 0.93$

Depreciation rate = 7%

EXERCISE 4E

1. A car is bought for $32000 and loses 8% of its value each year. What is the value after?
   (i) 2 years
   (ii) 9 years
   (iii) n years

2. A machine is bought for $8000 and loses 13% of its value each year. What is the value after:
   (i) 3 years
   (ii) 5 years
   (iii) n years

3. A PSP is bought for $1200 and loses 3.5% of its value each year. What is the value after 3 years?

4. An item is bought for $12000 and is valued at $8500 after 4 years. Find the constant rate of depreciation that would allow this drop in value.

5. An item is bought for $10000 and is valued at $7000 after 5 years. Find the constant rate of depreciation that would allow this drop in value.

6. An item halves its value after 3 years. Find the constant rate of depreciation that would allow this drop in value.

CHAPTER 5

FINANCIAL MATHEMATICS

5A   CALCULATING WAGES FROM AN ANNUAL SALARY

In this section, the reader's objective is to work out the weekly, fortnightly or monthly wage given the annual salary. What is annual salary? Annual salary is a fixed amount of money or compensation paid to an employee by an employer in return for any kind of work performed over a period of one year. In simple words, annual salary is the amount a person is paid during the period of a year (12 months, 52 weeks, 26 fortnights).

EXAMPLES

1. Peter earns $48 000 annually. Calculate his monthly wage. **SOLUTION** Monthly wage = $\frac{48000}{12}$ = $4000.	2. John gets an annual salary of $62 500. What is his weekly wage? **SOLUTION** Weekly wage = $\frac{62500}{52}$ = $1201.92

EXERCISE 5A

1. The table below shows the income earned by 4 friends working for Beta Chemicals. Complete the table.

	Annual income ($)	Monthly income ($)	Weekly income ($)	Fortnightly income ($)
Alex	42000			
Bob		3125		
Carl			745	
Emma				968

2. Calculate the weekly income for each professional given their annual salary.

Head Teacher Salary $96 000	Football player Salary $248 000	Scientist Salary $86 500

3. Manuela earns a salary of $36460 per annum. If she works 42 hours in a week, how much does she get paid for that week?

4. Maria's job pays her $1650 a month. What is her gross weekly wage?

5. John grosses $29400 a year. What is his gross weekly wage?

6. Sarah is a teacher and earns $48000 a year. Calculate her fortnightly gross pay.

7. Kevin works for ABC radio as an engineering tech and earns $45600 a year. Calculate his weekly gross pay.

8. Jane works a 36 hour week and earns an annual salary of $48600. Her friend Anna works part time and earns $23.80 per hour.

   a How much does Jane earn each week?

   b Who has the higher hourly rate of pay?

   c If Anna works on average 24 hours per week, what is her yearly income?

5B CALCULATING WAGES FROM AN HOURLY RATE

Wage is basically how much a person is paid to do a job. It can be measured weekly, monthly or annually.

Hourly rate, on the other hand, is simply how much a person gets paid for an hour of work.

EXAMPLE 1
Luke works at Mc Donald and his hourly rate of pay is $ 11.50. Last week he worked for 12 hours. Calculate his basic pay.

SOLUTION
Wage = 11.50 × 12 = $ 138

EXAMPLE 2
Olivia starts work at 7.30am and finishes at 4 pm. She had a lunch break between 12 pm and 1 pm. Calculate her basic pay if she earns $ 10.75 per hour of work.

SOLUTION

```
        30 mins            4 hours                    3 hours
7.30 am  ──→  8.00 am  ──→  12 pm    1pm  ──→  4 pm
```

Total hours worked = 7.5 hours

Wage = 7.5 × 10.75 = $ 80.63

EXERCISE 5B

1. Peter works at Hungry Jacks and his hourly rate of pay is $ 10.80. Last week he worked for 14 hours. Calculate his basic pay.	2. Becky starts work at 8.30 am and finishes at 3 pm. She had a lunch break between 12 pm and 12.30 pm. Calculate her basic pay if she earns $ 9.80 per hour of work.
3. Shania drives a truck for $18.25 an hour and works 38 hours a week. What is her annual gross pay?	4. Penny works as a waitress at the Cheese Cake Factory and makes $10.40/hour plus tips. Last week she worked 32 hours and made $463 in tips. What was her gross pay for the week?

5. Richard is a word processor operator. He makes $14.50 an hour and works 32 hours a week. What would be his fortnightly gross pay?

6. Anastasia works a basic week of 40 hours and her basic rate is $9.80 per hour. Calculate her basic wage for the week.

7. Jeremy works as a waiter in a Japanese restaurant. In addition to his regular pay of $11.20/hour, Jeremy keeps 85% of all the tips he receives. Calculate his gross weekly pay for a week in which he works 36 hours and receives $260 in tips.

8. Ryan Biggs work full time four days a week at Target and the table below shows his Week 13 work schedule. He earns $12.40 per hour.

Name : Ryan Biggs			Employee N0. 1831		Week 13
	In	Out	In	Out	Total hours worked
Mon	0900	1200	1300	1600	
Tue	0800	1200	1400	1700	
Wed	1030	1230	1300	1530	
Fri	0800	1200	1400	1930	
				Total	

(a) Complete the table, stating the total hours worked each day.

(b) Hence, calculate his wage for the whole week.

5C   OVERTIME AND OTHER ALLOWANCES

Overtime refers to doing extra work over and above your normal hours. Overtime allows people to earn a better hourly rate and thus more income. In this part of the chapter, emphasis will be laid on the two most common rates of overtime : double time (×2) and time and a half (× 1.5).

EXAMPLES

1. Paul the junior plumber works for $ 22 per hour. His overtime rate is "double time". What does he get paid for 3 hours overtime?
   SOLUTION
   3 hours overtime = 3 × 22 × 2 = $ 132.

2. Jacob the painter works a basic 40 hours a week. He does 2 hours overtime at 'time and a half' on Saturday and 4 hours 'double time' on Sunday. His hourly rate is $ 18 per hour. Work out his total pay for the week.
   SOLUTION
   2 hours' time and a half
   = 2 × 18 × 1.5 = $ 54
   4 hours double
   = 4× 18 × 2 = $ 144
   40 hours basic time
   = 40 × 18 = $ 720
   Total pay for the week
   = 54 + 144 + 720 = $ 918

3. Sam is employed at a hairdressing salon. He receives $16.50 per hour for a standard 35-hour Monday to Friday week. Sam also receives $45 per week as travel allowance and $200 per year as laundry allowance. Calculate Sam earnings for a standard week, assuming he receives his laundry allowance on a weekly basis.
   SOLUTION
   35 standard week = 35 × 16.50 = $ 577.50
   Travel allowance = $45
   Laundry allowance = 200 ÷ 52 = 3.85
   Total earnings per week = 577.50 + 45 + 3.85 = $ 626.35

4. Courtney is paid time and a half for each hour she works over 32 hours in a week. Last week she worked 40 hours for a total of $726. What is her normal hourly rate?
   SOLUTION
   She worked 8 hours overtime
   This is equivalent to 8 × 1.5 = 12 *hours* of normal hourly rate work
   She worked 32 + 12 = 44 *hours* at the normal hourly rate
   Normal hourly rate = $\frac{726}{44}$ = $16.50

5. A secretary works a 35-hour week for which she is paid $444.50. In a particular week she works 4 hours overtime on Saturdays which is paid for at time-and-a-half, and 2.5 hours overtime on Sunday which is paid for at double-time, calculate her gross wage for that week.
   SOLUTION
   Hourly rate = $\frac{444.50}{35}$ = $12.70
   Saturday's wage = 4 × 1.5 × 12.70 = $76.20
   Sunday's wage = 2.5 × 2 × 12.70 = $63.50
   Gross wage = 444.50 + 76.20 + 63.50 = $584.20

EXERCISE 5C

1. Peter the trainee electrician works for $24.50 per hour. His overtime rate is time-and-a-half. What does he get paid for 4 hours overtime?	2. Jimmy works for $18.50 per hour. His overtime rate is "double time". What does he get paid for 5 hours overtime?
3. A machinist is paid $21.60 per hour during normal working hours (9 am – 4 pm). For each hour after 4 pm, he is paid time-and-a-half. Calculate the amount received by the machinist on a day when he worked 9 am until 6 pm.	4. Alex is paid double time for each hour he works over 36 hours a week. Last week he worked 41 hours and earned $533.60. Calculate his normal hourly rate.
5. Calculate how much Thomas earns if he works 25 hours at the normal rate of $16.20 per hour, 3 hours at time-and-a-half, and 4 hours at double time.	6. In a week, Sammy works 15 hours at the normal rate, 4 hours at time-and-a-half, and 2 hours at double time. What is his hourly rate if he earns $285?
7. Harry is paid $18 per hour for a 38 hour week. If he works 45 hours in a week with overtime being paid time-and-a-half, how much does he earn for the week?	8. Gary is paid $16.50 per hour for a 40 hour week. If he works 51 hours in a week with 5 hours being paid time-and-a-half and the rest double-time, how much does he earn for the week?

5D PIECEWORK

Piecework refers to any type of job where a worker is paid a fixed rate for each unit produced irrespective of time taken. Many companies use the piecewise strategy to encourage efficiency and to increase productivity. However, there are other costs associated with piecework such as Quality Control.

EXAMPLES

1. Jay's Construction Company makes sandbags. Jay gets paid $0.32 a bag. He makes 158 bags a day, six days a week. What is his gross pay? **SOLUTION** Gross pay = 158 × 6 × 0.32 = $ 303.36	2. Peter works five days a week for a plastic company that pays him for each unit he completes. He receives 70 cents a unit for the first 80 units he produces each day and 75 cents a unit for all units over 80. Find his weekly income if he produces 105 units on average daily. **SOLUTION** Daily income = 80 × 0.70 + (105 − 80) × 0.75 = $ 74.75 Weekly income = 5 × 74.75 = $ 373.75

EXERCISE 5D

1. Peppa Sweet Company makes lolly bags. George gets paid $0.18 a bag. He makes 244 bags a day, five days a week. What is George's gross pay?	2. Samantha has a part-time job at Christmas, working for Australia Post. She receives $2.50 for each package she delivers. She delivers 24 packages on average in an evening. What will her gross pay be if she works four nights a week?
3. Gary works six days a week for a garment company that pays him for the units he completes. He receives 75 cents a unit for the first 60 units he produces each day, 80 cents per unit for the next 40, and 90 cents a unit for all units over 100. Find his weekly income if he produces 110 units on average daily.	4. Sarah makes socks at the local factory. She is paid 65 cents for every pair of socks she makes. If she made 1672 pair of socks this week, what is Sarah's gross pay?

5. Catherine hand paints wall plaques. She is paid $0.85 for every plaque she paints. Catherine painted 495 plaques this week. What is her gross pay?

6. Kat delivers magazines for a local restaurant. She is paid $0.32 per magazine she delivers. Kat is paid weekly. If she delivered 1452 magazines this week, what is her gross pay?

7. Terri works on her parents grape vines. She is paid $30 plus $0.85 for each basket she fills. If she managed to fill 65 baskets, how much did she earn?

8. Anthony delivers phone books in his suburb. He is paid $16.25 plus 23 cents for each phone book. If he delivered 280 phone books, how much did he make?

9. Leonardo is a finisher in a furniture plant. He is paid $1.40 for every coffee table he finishes and $1.05 for every end table he finishes. Leonardo finished 198 coffee tables and 395 end tables this week. What is Leonardo's gross pay for the week?

10. Alexa assembles porcelain dolls at a factory. She is responsible for putting on the legs. She is paid $32 per day, plus 18 cents for each pair of legs she attaches. Yesterday, she attached 398 pairs of legs. How much did she earn?

11. Joshua is a cheese packer at Long River Cheese Factory. He works every day except Saturday. He is paid $0.48 for every case of cheese he packs and loads onto delivery pallets Monday through Friday and $0.63 for every case he packs and loads on Sunday. Joshua packed and loaded the following cases of cheese this week:

Monday	Tuesday	Wednesday	Thursday	Friday	Sunday
235	245	210	285	190	312

What is Joshua's gross pay for the week?

5E   EXCHANGE RATES

The rate of exchange between different currencies varies on a daily basis depending on lot of factors such as economic stability, political situation, social factors and so on.

The table below shows the exchange rates of one Australian dollar (AUD) vis-à-vis other currencies as at 10th July 2014.

TABLE 1		
Currency	**Symbol**	**Exchange Rate**
US Dollar (USD)	$	0.94
British Pound Sterling (GBP)	£	0.55
Chinese Yuan (CNY)	¥	5.82
Euro (EUR)	€	0.69
Hong Kong Dollar (HKD)	HK$	7.26
Japanese Yen (JPY)	¥	95.11
New Zealand Dollar (NZD)	$	1.06
South African Rand (ZAR)	R	10.05
Swiss Franc (CHF)	CHF	0.84
Indonesian Rupiah (IDR)	Rp	10841
Thai Baht (THB)	฿	30.17

EXAMPLES

1. Matthew changes 25 200 Japanese Yen into Australian dollars.
   Calculate how many dollars he receives.
   SOLUTION
   $$\frac{25200}{95.11} = \$264.96$$

3. Eddie travels from Australia to Hong Kong. He changes $450 into Hong Kong Dollar.

   (a) How many Hong Kong Dollar does he receive?

   $$450 \times 7.26 = \$3267$$

   (b) When Eddie returns to Australia he has 81 HKD. How many $ AUD does he get in return?

   $$\frac{81}{7.26} = \$11.16$$

2. Alex exchanges New Zealand Dollars into Australian Dollar. Bob exchanges British Pound Sterling into Australian Dollar as well. Given that both received $250 (AUD) in return for their exchanges, determine how many NZD did Alex change and the number of GBP Bob converted.
   SOLUTION
   Alex : $250 \times 1.06 = \$265 \; NZD$

   Bob : $250 \times 0.55 = 137.50 \; GBP$

EXERCISE 5E

1. Complete the following table.

Currency	Exchange Rate to $AUD	Amount converted	Australian Dollars received
US Dollar (USD)	0.94	320	
British Pound Sterling (GBP)	0.55	160	
Chinese Yuan (CNY)	5.82	4200	
Euro (EUR)	0.69	240	
Hong Kong Dollar (HKD)	7.26	600	
Japanese Yen (JPY)	95.11	45 000	
New Zealand Dollar (NZD)	1.06	680	
South African Rand (ZAR)	10.05	860	
Swiss Franc (CHF)	0.84	450	
Indonesian Rupiah (IDR)	10841	2 000 000	
Thai Baht (THB)	30.17	60 000	

2. Convert the following Australian dollars to the different currencies.

Amount of Australian dollars converted	Currency converted to	Exchange Rate	Amount received
4600	US Dollar (USD)	0.94	
150	British Pound Sterling (GBP)	0.55	
5700	Chinese Yuan (CNY)	5.82	
320	Euro (EUR)	0.69	
800	Hong Kong Dollar (HKD)	7.26	
700	Japanese Yen (JPY)	95.11	
680	New Zealand Dollar (NZD)	1.06	
860	South African Rand (ZAR)	10.05	

3. On one particular day the rate of exchange between pounds (£) and Australian dollars ($) was £1 = $1.85.
   Calculate
   (a) the number of dollars received in exchange for £150,

   (b) the number of pounds received in exchange for $264.

   (c) John buys 24 postcards for £1.30 each. Calculate the total cost, in dollars, of the postcards.

4. The exchange rate between pounds and Australian dollars is £1 = $1.76.
   Anna converts $280 into pounds.
   Calculate the number of pounds Anna receives.

3. The exchange rate between Australian dollars and euros is $1 = €0.85.
   Ben changes $260 into euros.
   Calculate the number of euros Ben receives.

6. Five items bought at Freddo's Supermarket are shown on the receipt.

   The part showing the cost of the apples is missing.

   (a) How much did the apples cost?

   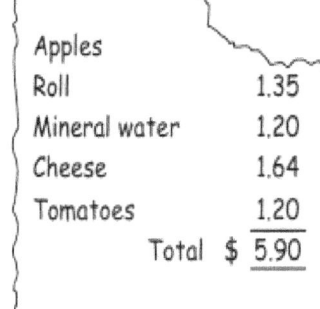

   (b) The total cost of $5.90 when converted to euros is €4.80.
   Determine the exchange rate that will enable you to convert dollars ($) to euros (€).

   (c) Use your answer to (b) to estimate the cost of cheese in euros.

5F BEST BUYS

CLASS ACTIVITY

1. Complete the following table

	Cost per gram	Cost per 100g	Cost per kg
$4 for 50g			
$15.30 for 250g			
$12.60 for 400g			
$24.60 for 1 kg			
$60 for 2.5kg			

2. Two shops, Food Mart and Jim's Store, both sell Sweet Yoghurts.

Seven Eleven	P Mart
4 for $5.80	5 for $7

At which shop are Sweet Yoghurts the better value for money? Show all your working.

3. Tomatoes cost $16.50 for a 12.5 kg bag at a farm shop. The same type of tomatoes cost $3.75 for a 3 kg bag at a supermarket. Where are the tomatoes the better value, at the farm shop or at the supermarket?

4. Sprite is on offer at Moles and Foodworth. Moles : 2 litre bottles on offer 3 for $9.30 Foodworth: 24 cans x 330ml on offer for $24.35. Where is Sprite on a better deal?

5G BUDGETING

Budgeting is often described as an estimate of expected income and expenditures for a given period in the future. It is sometimes itemised and the goal of budgeting is to set a total sum of money aside needed for a purpose: planning a holiday during the school breaks. Budgeting also refers to subsist on or live within a particular budget.

FIXED EXPENDITURE V/S DISCRETIONARY SPENDING

A fixed expense is a cost or obligation that occurs regularly and doesn't vary in amount. Examples of Fixed Expenditures can be Rent or Insurance.

A discretionary spending, on the other hand, is an amount of a person's income that is left for spending after personal necessities such as food, shelter, and clothing have been paid.

CLASS ACTIVITY

Categorise the expenses in the table below as fixed or discretionary.

Description	Fixed	Discretionary	Description	Fixed	Discretionary
Mortgage			Groceries		
Restaurants			Music Lesson fee		
Vacation			Utilities		
Massage			Car payment		
Gym Subscription			Pets		

CLASS ACTIVITY 2

Make a list of the income you receive and the expenses you have each month and prepare a budget for yourself. If your income and expenses are equal, you aren't saving anything. Do you think this is a problem? If your income exceeds your expenditure, do you find it possible to save to undertake a project such as buying a car, travel etc.. Explain your thinking.

EXERCISE 5G

1. The following table shows the income and expenditure for the Williams family for one year.

Income ($)		Expenditure ($)	
Salaries & Wages	28 240.00	Food	5400.00
Centre Link	6 350.64	Rent	9 600.00
		Power	3 453.60
		Insurance	950.55
		Telephone	1 456.14
		Other	6523.41
		Entertainment	4523.69
Total		**Total**	

(a) Complete the table by calculating the total income and expenditure.

(b) What percentage of their income did the Williams family save?

(c) The Williams family is planning for a holiday costing $12000. If they plan to set aside 75% of the savings for their holiday, how long would it take for the family to attain their goal?

(d) Split the expenditure items under the headings fixed expenditure or discretionary spending.

CHAPTER 5 : FINANCIAL MATHEMATICS

2. Jack and Kelly are both 14 years of age attending the same school. Their parents have agreed they could purchase a car in three years when they both get their licenses at 17. Their aim is to save enough over the next three years to buy the car themselves. They also have to pay for costs like gas, repairs, and insurance. Jack and Kelly found a reasonably priced car for $5000, an amount that they thought they could afford. Bearing inflation in mind the car might cost $5500 in 3 years' time. They decided to make a budget estimate of their expected income and expenditures. Their budget is tabulated below for each month.

Monthly Income & Expenditure	Jack	Kelly
Allowance	$75	$75
Games rental	$8	$0
Part-time job	$128	$116
Snacks	$21	$13
School supplies	$11	$19
Phone rental	$15	$15
Entertainment	$22	$15

(a) List the income and expenditure for both Jack and Kelly and calculate how much they save each month.

(b) After three years, will they have saved enough to afford the car?

(c) How much more money does each one need to save each month to afford to buy the car?

(d) Jack and Kelly are budgeting to make sure they save enough to buy the car. They also have to consider the expenses they will face to operate the car after they buy it. List a few operating expenses they might include.

3. The table below shows the charges associated with water usage in a Royal City.

Rates for reading the water meter	
Usage (KL) per year	Meters read January–December 2014
first 150 kL	85.2 c/kL
Next 200 kL	120.5 c/kL
next 150 kL	132.6 c/kL
over 500 kL	169.4 c/kL

(a) The account shows that the water usage for the Smith family was 230 kL. Calculate the amount they are required to pay for their water usage.

(b) The Bligh's water usage was 450 kL. Calculate the amount they are required to pay for their water usage.

(c) The Simpson's have a budget of $500 for their water usage. Calculate by how much they exceeded their budget if their water consumption was exactly 540 kL?

CHAPTER 5 : FINANCIAL MATHEMATICS

5H SHARES AND DIVIDENDS

Dividend is the distribution of a fraction of a company's earnings to its shareholders, usually decided by the board of directors. Dividend is often quoted in terms of the dollar amount each share receives also referred to as "Dividend Per Share (DPS)."

Dividends may be in the form of cash, stock or property. Most secure and stable companies offer dividends to their shareholders. High-growth companies rarely offer dividends because all of their profits are reinvested to help sustain higher-than-average growth.

EXAMPLE 1

Calculate the dividend paid on the portfolio of shares given the dividend paid for each share.

Company	Number of shares	Price per share	Dividend per share
AB Co Ltd	2000	$1.20	$0.15
DNDS Inc	450	$4.60	$0.60
Golden Miners	6000	$36.55	$2.40

SOLUTION

Dividend paid by AB Co Ltd = 2000 × 0.15 = $300

Dividend paid by DNDS Inc = 450 × 0.60 = $270

Dividend paid by Golden Miners = 6000 × 2.40 = $14400

Total dividend = $14970

EXAMPLE 2

Calculate the dividend paid on the portfolio of shares given the percentage dividend paid per share.

Company	Number of shares	Price per share	Percentage Dividend per share
Junior Ltd	5500	$8.60	5.6%
The Croods Inc	680	$15.40	3.9%
Philip Trading	4250	$3.75	No dividend

SOLUTION

Dividend paid by Junior Ltd = 5500 × 8.60 × 5.6% = $2648.80

Dividend paid by Croods Inc = 680 × 15.40 × 3.9% = $408.41

Dividend paid by Philip Trading = 4250 × 0 = $0

Total dividend = $3057.21

EXERCISE 5H

1. Calculate the dividend paid on the portfolio of shares given the dividend paid for each share.

Company	Number of shares	Price per share	Dividend per share
Avengers Ltd	800	$2.20	$0.25
Knight Inc	1250	$10.60	No Dividend
Thor Mining	630	$15.40	$1.70
Green lantern Ltd	5800	$5.80	$0.90

2. Calculate the dividend paid on the portfolio of shares given the percentage dividend paid per share.

Company	Number of shares	Price per share	Percentage Dividend per share
Alpha Ltd	650	$12.40	4.6%
Beta Corp	7000	$7.20	5.1%
Gamma Inc	980	$11.55	No dividend
Delta Mining	2500	$6.90	8.3%

3. Calculate the dividend paid on the portfolio of shares given the dividend paid for each share.

Company	Number of shares	Price per share	Dividend per share
Company A	3200	$2.90	$0.35
Company B	780	$11.35	$1.30
Company C	5120	$8.20	No dividend
Company D	2670	$6.40	$1.42
Company E	925	$16.60	$2.65

4. Calculate the dividend paid on the portfolio of shares given the percentage dividend paid per share.

Company	Number of shares	Price per share	Percentage Dividend per share
Murray Ltd	1230	$10.20	2.1%
Swan Corp	3500	$5.30	4.6%
Canning Inc	2800	$9.75	5.3%

5I   PRICE–TO–EARNINGS RATIO

Price-to-earnings ratio, commonly known as P/E ratio, is a tool that is used by investors deciding whether they should buy a particular stock or not. The P/E ratio implies how much an investor has to pay for every $1 of earnings. A low P/E ratio is attractive implying that one pays less for every $1 of earnings. Similarly, a company with a higher P/E ratio generally expect higher earnings growth in the future.

$$Price-to-Earnings\ ratio = \frac{Price\ per\ share}{Earnings\ per\ share}$$

Price per share in simple words is the cost of buying a share of any company on the stock Market. Earnings per share, on the other hand, is calculated by taking a company's net income over the last twelve months, subtracting any dividends, and then dividing the difference by the number of shareholders.

CLASS ACTIVITY

1. Calculate the P/E ratio for each of the following companies.

Company	Market value per share	Earnings per share	P/E Ratio
DK Trading	15.50	1.50	$\frac{15.50}{1.50} = 10.3$
RJ Mining	8.60	0.80	
ABC Banking	1.65	0.25	
Mars Bar Co Ltd	49.43	2.55	
Top Shop Ltd	12.60	2.40	

2. Complete the following table.

Company	Market value per share	Earnings per share	P/E Ratio
SOS Inc	18.60		7.75
Delta Trading	9.25	1.25	
Swiss Choc Factory	15.60		10.4
MBCL Ltd		1.20	19
Alien Groups	2.80	0.40	

3. Alan has the choice to buy shares in two different companies A and B. He collected some information about both companies as shown under.

	Company A	Company B
Price per share	$16	$15
EPS	$2	$1
P/E ratio	8	15

Which company would you advise Alan to invest in?

4. You have been assigned to evaluate the following three companies stocks within the same industry:

Company	Price per share	Earnings Per Share
Cybertrons Corporation	$52.40	$12.50
Decepticons Ltd	$8.50	$2.25
Optimus Prime Inc	$9.88	$5.25

Which stock has the best value? Show workings.

5J ALLOWANCES : FAMILY TAX BENEFIT

Family Tax Benefit (FTB) is an allowance that helps eligible families with the cost of raising their children. It is made up of two parts FTB Part A and FTB Part B. FTB Part A is paid depending on the number of children in a family and their financial status. FTB Part B, on the other hand, is paid per-family and gives extra help to single parents and families with one main income.

The maximum amounts of Family Tax Benefit Part A received per child are updated on 1 July each year. The table below shows the current rates per child within different age brackets and circumstances. For the scope of this book, only FTB part A has been included as examples.

TABLE 1 : CURRENT RATES FTB Part A	
For each child	Per fortnight
0 to 12 years	$169.68
13-15 years	$220.64
16-19 years	$220.64
16-17 years (completed secondary study)	$54.32
18-21 years (completed secondary study)	$54.32

To be eligible for the above rates, a particular family must undergo one of the following tests depending on their yearly income.

TABLE 2 : FTB Part A Tests	
TEST 1	Maximum rate for Family Tax Benefit Part A less 20 cents for each dollar above $50,151.
TEST 2	Maximum rate for Family Tax Benefit Part A less 30 cents for each dollar above $94, 316, plus $3,796 for each Family Tax Benefit child after the first.

EXAMPLES

1. The Smith family has two children aged 5 and 8 years. The combined family income is $63500. Calculate the amount of FTB Part A received by the Smith family.
SOLUTION
Yearly maximum amount for both children
$= 169.68 \times 2 \times 26 = \$8823.36$
Applying the first test :
$(63500 - 50151) \times 0.20 = \$2669.80$
FTB Part A received
$8823.36 - 2669.80 = \$6153.56$

2. The Gordon family has two children aged 10 and 17 years. The 17 year old son has completed his secondary study and works part-time in a restaurant. The family net income is $59860. Determine the amount of FTB Part A received by the Gordon family.
SOLUTION
Yearly maximum amount for both children
$= 169.68 \times 26 + 54.32 \times 26 = \$5824$
Applying the first test :
$(59860 - 50151) \times 0.20 = \$1941.80$
FTB Part A received
$5824 - 1941.80 = \$3882.20$

EXERCISE 5J

Referring to Tables 1 and 2 on the previous page, answer the following questions.

1. A family's income is $55765 a year. The family has two children of 3 and 9 years respectively. Calculate the amount of FTB Part A received by this family.	2. Mr and Mrs Pavilion have three children : a son aged 11 years and twin daughters of 15 years of age. The Pavilions' run their own business and earn $75420 a year. Calculate the amount of FTB Part A received by the family.
3. Calculate the amount of FTB Part A received by the Packard family having only one child aged 14 years and having a net annual income of $90000.	4. A family's income is $1985 a week. The family has two children of 7 and 16 years respectively. Calculate the amount of FTB Part A received by this family.

5K CARER ALLOWANCE AND OLD AGE PENSIONS

CARER ALLOWANCE : A carer allowance is an additional payment made to people who provide additional daily care and attention for someone with a disability or medical condition. Carer Allowance when caring for a child less than 16 years is currently a fortnightly payment of $118.20. The payments are however adjusted on 1st January each year to match the cost of living. Note that Carer Allowance is a non-taxable payment.

AGE PENSIONS : In order to be eligible for Age Pensions in Australia, a person has to meet the residence requirements. If you are an Australian citizen and have been residing in Australia for at least ten years cumulatively then you can be eligible for the age pension. Permanent residents living in Australia may also be eligible for age pension.

In addition to residence requirements, there's also an assets test and income test the Government will carry out to assess whether a particular person is eligible for the Age Pension or not. The tests take into account factors such as age pension, wife pension, carer payment and so on.

The table below shows the Pension rates for Age Pensions.

TABLE 1		
Pension rates for Age Pensions		
Pension rates per fortnight	Single	Couple each
Maximum basic rate	$776.70	$585.50

The **ASSETS TEST** has two thresholds and is split into two categories as shown in the table below.

		TABLE 2		
		THE ASSETS TEST		
	Home Owners		**Non-Home Owners**	
	Lower Threshold	Upper threshold	Lower Threshold	Upper threshold
Single	192 500	707 750	332 000	847 250
Couples	273 000	1 050 000	412 500	1 189 500

The full age pension is received when a lower assets test threshold is not exceeded. Once the lower thresholds are exceeded a person or couple's entitlement to the age pension is reduced by $1.50 a fortnight for every $1000 their assets exceed that threshold. Unfortunately, no age pension is received once an upper threshold is exceeded.

The table below shows the **INCOME TEST** for singles and couples.

	TABLE 3	
	THE INCOME TEST	
	Payment per fortnight	**Reductions**
Single	Up to $160	None – full payment
	Above $160	50 cents for each dollar over $160
Couples combined	Up to $284	None – full payment
	Above $284	50 cents for each dollar over $284

For a person being single and earning income up to $160 a fortnight, there is no deduction in his or her pension. However, for income exceeding $160, there is a deduction of 50 cents per dollar.

EXAMPLES

1. Using the information from Table 1 and Table 2 on the previous page, determine the fortnightly age pension of a single non-homeowner of pension age owning assets worth $390 000.
   SOLUTION

   Fortnightly Age pension

   $= 776.70 - \left(\dfrac{390000 - 332000}{1000}\right) \times 1.50$
   $= \$689.70$

2. Mr and Mrs Jones, both in their late sixties, have invested a sum of $200 000 which earns them a yearly simple interest of 5%. Calculate their combined fortnightly age pension using Table 1 and Table 3.

   SOLUTION

   $Interest\ earned = 200000 \times 0.05$
   $= \$10000$
   Fortnightly income $= 10000 \div 26 \approx \$385$
   Fortnightly Age pension

   $= 585.50 \times 2 - (385 - 284) \times 0.50$
   $= \$1120.50$

EXERCISE 5K

1. Using the information from Table 1 and Table 2 on the previous page, determine the fortnightly age pension of a single homeowner of pension age owning assets worth $405 000.

2. Mr Adan Bligh, a widower in his early seventies, has invested a sum of $150 000 which earns him a flat fixed rate of 6% each year. Calculate his fortnightly age pension using Table 1 and Table 3.

3. Using the information from Table 1 and Table 2 on the previous page, determine the fortnightly age pension of a home owning couple both of pension age owning assets worth $580 000.

4. Mrs Blitz is a 69 year old widow. She has invested $90 000 as a shareholder in an IT company. Last year she received 2% flat interest on her investment plus $450 as dividend each fortnight. Using Tables 1 and 3, calculate her fortnightly age pension.

5. Agent Coulson is a single homeowner and a retired army officer and has reached the pension age. He has assets worth $365 000. He has invested $165 000 of his assets in Delta Bank earning 4% per annum as interest.

   (a) Determine using the Assets Test his fortnightly pension.

   (b) Determine using the Income Test his fortnightly age pension.

   (c) It is customary for the government to pay the lower pension out of the Income test or Assets test carried out. State Agent Coulson's fortnightly pension rounded up to the nearest dollar.

CHAPTER 6

MATRICES

A matrix is an array of numbers arranged in rows (horizontally) or in columns (vertically) or both. Consider the following example.

A café sells tea, juice and coffee, each in small and large cups.

The cost of a small cup is $3.50 for all three types of drinks and the cost for large cup is $4.25.

The table below shows the number of cups sold during a period of one hour.

Drink	Small	Large
Tea	9	5
Juice	12	3
Coffee	7	8

The above table can be represented in a matrix as shown.

$$\begin{bmatrix} 9 & 5 \\ 12 & 3 \\ 7 & 8 \end{bmatrix} \rightarrow three\ rows$$

2 columns

The numbers represented horizontally are the rows and the numbers written vertically are called columns. In the above matrix there 3 rows and 2 columns. Each of the numbers in the matrix are called the elements of the matrix and are denoted as e_{ij}, i being the row number and j being the column number. For example, $e_{22} = 3$.

As a convention, we use capital letters and square brackets to represent a matrix.

CLASS ACTIVITY 1

$\begin{bmatrix} 4 & 2 & 5 & 8 \\ -1 & 0 & 3 & -7 \\ 3 & 7 & 9 & 1 \end{bmatrix}$	$e_{21} = -1$ $e_{33} = 9$ $e_{14} = 8$	$\begin{bmatrix} 4 & 0 \\ 1 & 2 \\ 3 & 4 \\ 5 & 6 \end{bmatrix}$	$e_{21} =$ $e_{32} =$ $e_{42} =$
$\begin{bmatrix} 5 & 1 & -4 \\ 0 & 8 & 6 \\ 4 & 2 & 7 \end{bmatrix}$	$e_{31} =$ $e_{23} =$ $e_{12} =$	$\begin{bmatrix} 5 & -2 & -5 \\ 1 & 3 & -3 \\ 2 & 4 & 0 \\ 7 & 8 & 9 \end{bmatrix}$	$e_{23} =$ $e_{32} =$ $e_{41} =$

6A SIZE (ORDER) OF A MATRIX

Consider the matrix $\begin{bmatrix} 9 & 5 \\ 12 & 3 \\ 7 & 8 \end{bmatrix}$. Since it has three rows and two columns, we say the order of the matrix is 3 × 2 (read as 3 by 2). Therefore,

Order of a matrix = number of rows (r) × number of columns (c).

CLASS ACTIVITY 2

Complete the following table, where necessary.

MATRIX	SIZE	MATRIX	SIZE
$\begin{bmatrix} 5 & 2 & -1 \\ 6 & 4 & 7 \end{bmatrix}$	2 × 3	$\begin{bmatrix} 6 \\ -2 \\ 5 \end{bmatrix}$	3 × 1
$\begin{bmatrix} 1 & 7 \\ 5 & 4 \end{bmatrix}$		$\begin{bmatrix} 4 & 8 \\ 0 & 1 \\ 7 & 3 \end{bmatrix}$	
$\begin{bmatrix} 7 & 11 \end{bmatrix}$		$\begin{bmatrix} 4 & 0 \\ 1 & 2 \\ 3 & 4 \\ 5 & 6 \end{bmatrix}$	
$\begin{bmatrix} 5 \\ -1 \\ 3 \\ 8 \end{bmatrix}$		$\begin{bmatrix} 11 \end{bmatrix}$	
$\begin{bmatrix} 0 & 1 & 2 \end{bmatrix}$		$\begin{bmatrix} 5 & 1 & 0 \\ 0 & 8 & 6 \\ 4 & 2 & 7 \end{bmatrix}$	
$\begin{bmatrix} 4 & 2 & 5 & 2 \\ -1 & 0 & 3 & 7 \\ 3 & 7 & 9 & 1 \end{bmatrix}$		$\begin{bmatrix} 1 \\ 2 \\ 4 \\ 7 \\ 3 \end{bmatrix}$	

6B   TYPES OF MATRICES

There are several types of matrices but in this section emphasis will be laid on row, column, square, zero, diagonal, identity and equal matrices.

ROW MATRIX

It is a matrix having only one row. The size of a row matrix is in the form $1 \times c$, where c represents the number of columns.

Examples of row matrices are $P = \begin{bmatrix} 0 & 1 & 2 \end{bmatrix}$ or $Q = \begin{bmatrix} 5 & 2 & 4 & 7 \end{bmatrix}$.

COLUMN MATRIX

It is a matrix having only one column. Since there is only one column the size of a column matrix is always written in the form $r \times 1$, where r represents the number of rows.

Examples of column matrices are : $A = \begin{bmatrix} 6 \\ -2 \\ 5 \end{bmatrix}$ or $B = \begin{bmatrix} 1 \\ 2 \\ 4 \\ 7 \\ 3 \end{bmatrix}$

SQUARE MATRIX

It is matrix in which the number of rows is equal to the number of columns. Examples of square matrices are $C = \begin{bmatrix} 1 & 7 \\ 5 & 4 \end{bmatrix}$ or $D = \begin{bmatrix} 5 & 1 & 0 \\ 0 & 8 & 6 \\ 4 & 2 & 7 \end{bmatrix}$ or $E = \begin{bmatrix} 7 \end{bmatrix}$

ZERO MATRIX (NULL MATRIX)

It is a matrix in which all the elements are zeroes.

Examples of zero matrices are $M = \begin{bmatrix} 0 & 0 \\ 0 & 0 \end{bmatrix}$ or $N = \begin{bmatrix} 0 & 0 & 0 \\ 0 & 0 & 0 \\ 0 & 0 & 0 \end{bmatrix}$

DIAGONAL MATRIX

It is a matrix in which all the elements are zeroes with the exception of those elements lying in the leading diagonal; the latter slopes downwards from left to right.

Examples of diagonal matrices are $M = \begin{bmatrix} 1 & 0 \\ 0 & 3 \end{bmatrix}$ or $N = \begin{bmatrix} 4 & 0 & 0 \\ 0 & 5 & 0 \\ 0 & 0 & -2 \end{bmatrix}$

IDENTITY MATRIX (UNIT MATRIX)

It is a diagonal matrix in which all the elements are ones. We use the letter I to denote an identity matrix.

Examples of identity matrices are $M = \begin{bmatrix} 1 & 0 \\ 0 & 1 \end{bmatrix}$ or $N = \begin{bmatrix} 1 & 0 & 0 \\ 0 & 1 & 0 \\ 0 & 0 & 1 \end{bmatrix}$

EQUAL MATRICES

Two matrices are said to be equal if

(a) They have the same size
(b) Their corresponding elements are equal.

For example if $\begin{bmatrix} x & 0 \\ 2 & 3 \end{bmatrix} = \begin{bmatrix} 1 & 0 \\ 2 & y \end{bmatrix}$, then $x = 1$ and $y = 3$

CLASS ACTIVITY 3

Name the following matrices.

MATRIX	NAME	MATRIX	NAME
$\begin{bmatrix} 5 & 0 & 0 \\ 0 & 8 & 0 \\ 0 & 0 & 7 \end{bmatrix}$	Square diagonal	$\begin{bmatrix} 6 \\ -2 \\ 5 \end{bmatrix}$	
$\begin{bmatrix} 1 & 7 \\ 5 & 4 \end{bmatrix}$		$\begin{bmatrix} 4 & 8 \\ 0 & 1 \\ 7 & 3 \end{bmatrix}$	
$\begin{bmatrix} 7 & 11 \end{bmatrix}$		$\begin{bmatrix} 4 & 0 \\ 1 & 2 \\ 3 & 4 \\ 5 & 6 \end{bmatrix}$	
$\begin{bmatrix} 1 & 0 \\ 0 & 1 \end{bmatrix}$		$[11]$	
$\begin{bmatrix} 0 & 0 & 0 \end{bmatrix}$		$\begin{bmatrix} 0 \\ 0 \end{bmatrix}$	
$\begin{bmatrix} 2 & 3 & 4 \\ 1 & -8 & 3 \\ 4 & 0 & -3 \end{bmatrix}$		$\begin{bmatrix} 0 & 1 \\ 1 & 0 \end{bmatrix}$	

6C   ADDITION OF MATRICES

Matrices can be added provided they have the same size. Adding the corresponding elements in each matrix performs addition.

EXAMPLE

Given $A = \begin{bmatrix} 1 & 7 \\ 5 & 4 \end{bmatrix}$ $B = \begin{bmatrix} 3 & 5 \\ -2 & 0 \end{bmatrix}$ $C = \begin{bmatrix} 10 \\ 15 \end{bmatrix}$,

express as a single matrix.

(a) A + B   (b) B + C

SOLUTION

(a) $A + B = \begin{bmatrix} 1 & 7 \\ 5 & 4 \end{bmatrix} + \begin{bmatrix} 3 & 5 \\ -2 & 0 \end{bmatrix} = \begin{bmatrix} 4 & 12 \\ 3 & 4 \end{bmatrix}$

(b) $B + C = \begin{bmatrix} 3 & 5 \\ -2 & 0 \end{bmatrix} + \begin{bmatrix} 10 \\ 15 \end{bmatrix} =$ Not possible as they have different size.

EXERCISE 6C

Express the following as single matrices.

1.	$\begin{bmatrix} 5 \\ 3 \end{bmatrix} + \begin{bmatrix} 4 \\ -2 \end{bmatrix}$	2.	$\begin{bmatrix} 2 & 7 \\ 8 & 0 \end{bmatrix} + \begin{bmatrix} 8 & -1 \\ -2 & 0 \end{bmatrix}$
3.	$\begin{bmatrix} 5 & 1 & 0 \\ 0 & 8 & 6 \\ 4 & 2 & 7 \end{bmatrix} + \begin{bmatrix} 2 & 3 & 4 \\ 1 & -8 & 3 \\ 4 & 0 & -3 \end{bmatrix}$	4.	$\begin{bmatrix} 6 \\ -2 \\ 5 \end{bmatrix} + \begin{bmatrix} 4 & 8 \\ 0 & 1 \\ 7 & 3 \end{bmatrix}$
5.	$\begin{bmatrix} 4 & 3 \\ 6 & 5 \end{bmatrix} + \begin{bmatrix} 6 & -3 \\ -1 & 8 \end{bmatrix}$	6.	$\begin{bmatrix} -6 \\ 3 \end{bmatrix} + \begin{bmatrix} 10 \\ -7 \end{bmatrix}$
7.	$\begin{bmatrix} 2a & x \\ 8 & 3b \end{bmatrix} + \begin{bmatrix} 8a & 4x \\ 2 & 0 \end{bmatrix}$	8.	$\begin{bmatrix} 6 \\ -2 \\ 5 \end{bmatrix} + \begin{bmatrix} 2 \\ 0 \\ 3 \end{bmatrix} + \begin{bmatrix} 1 \\ -3 \\ -4 \end{bmatrix}$

6D SUBTRACTION OF MATRICES

Similar to addition, matrices can be subtracted provided they have the same size. Subtracting the corresponding elements in each matrix performs subtraction.

EXAMPLE

Given $P = \begin{bmatrix} 4 & -1 \\ 0 & 5 \end{bmatrix}$ $Q = \begin{bmatrix} 3 & 8 \\ -2 & -2 \end{bmatrix}$ $R = \begin{bmatrix} 9 \\ 11 \end{bmatrix}$,

express as a single matrix.

(a) P – Q   (b) Q – R

SOLUTION

(a) $P - Q = \begin{bmatrix} 4 & -1 \\ 0 & 5 \end{bmatrix} - \begin{bmatrix} 3 & 8 \\ -2 & -2 \end{bmatrix} = \begin{bmatrix} 1 & -9 \\ 2 & 7 \end{bmatrix}$

(b) $Q - R = \begin{bmatrix} 3 & 8 \\ -2 & -2 \end{bmatrix} - \begin{bmatrix} 9 \\ 11 \end{bmatrix}$ = Not possible as they have different size.

EXERCISE 6D

Express the following as single matrices.

1.	$\begin{bmatrix} 5 \\ 3 \end{bmatrix} - \begin{bmatrix} 2 \\ -2 \end{bmatrix}$	2.	$\begin{bmatrix} 2 & 9 \\ 5 & 0 \end{bmatrix} - \begin{bmatrix} 8 & -1 \\ -2 & 0 \end{bmatrix}$
3.	$\begin{bmatrix} 5 & 1 & 0 \\ 0 & 8 & 6 \\ 4 & 2 & 7 \end{bmatrix} - \begin{bmatrix} 4 & 8 \\ 0 & 1 \\ 7 & 3 \end{bmatrix}$	4.	$\begin{bmatrix} 6 \\ -2 \\ 5 \end{bmatrix} - \begin{bmatrix} 4 & 5 \\ 0 & -2 \\ 7 & 3 \end{bmatrix}$
5.	$\begin{bmatrix} 4 & 3 \\ 6 & 5 \end{bmatrix} - \begin{bmatrix} 6 & 4 \\ -1 & 7 \end{bmatrix}$	6.	$\begin{bmatrix} -6 \\ 3 \end{bmatrix} - \begin{bmatrix} 8 \\ -5 \end{bmatrix}$
7.	$\begin{bmatrix} 12a & 2x \\ 8 & 5b \end{bmatrix} - \begin{bmatrix} 7a & x \\ -2 & 7b \end{bmatrix}$	8.	$\begin{bmatrix} 6 \\ -2 \\ 5 \end{bmatrix} + \begin{bmatrix} 2 \\ 0 \\ 3 \end{bmatrix} - \begin{bmatrix} 1 \\ -3 \\ -4 \end{bmatrix}$

CHAPTER 6 : MATRICES

6E MULTIPLYING A MATRIX BY A SCALAR

A scalar is in fact just a random number. To multiply a matrix by a scalar, multiply all the elements of the matrix by the scalar as shown in the examples below.

EXAMPLE 1

Given $\quad A = \begin{bmatrix} 1 & 7 \\ 5 & 4 \end{bmatrix} \quad\quad B = \begin{bmatrix} 3 & 5 \\ -2 & 0 \end{bmatrix}$

Express as a single matrix.

(a) 3A (b) A + 2B

SOLUTION

(a) $3A = 3 \begin{bmatrix} 1 & 7 \\ 5 & 4 \end{bmatrix} = \begin{bmatrix} 3 & 21 \\ 15 & 12 \end{bmatrix}$

(b) $A + 2B = \begin{bmatrix} 1 & 7 \\ 5 & 4 \end{bmatrix} + 2 \begin{bmatrix} 3 & 5 \\ -2 & 0 \end{bmatrix}$

$\quad\quad\quad\quad = \begin{bmatrix} 1 & 7 \\ 5 & 4 \end{bmatrix} + \begin{bmatrix} 6 & 10 \\ -4 & 0 \end{bmatrix}$

$\quad\quad\quad\quad = \begin{bmatrix} 7 & 17 \\ 1 & 4 \end{bmatrix}$

EXAMPLE 2

If $3N = \begin{bmatrix} 6 & 12 \\ -3 & 0 \end{bmatrix}$, find N.

SOLUTION

$N = \frac{1}{3} \begin{bmatrix} 6 & 12 \\ -3 & 0 \end{bmatrix} = \begin{bmatrix} 2 & 4 \\ -1 & 0 \end{bmatrix}$

EXAMPLE 3

Find the matrix M which is such that $2M - \begin{bmatrix} 0 & 4 \\ -6 & 8 \end{bmatrix} = 3 \begin{bmatrix} 2 & 0 \\ 0 & -4 \end{bmatrix}$

SOLUTION

$$2M = \begin{bmatrix} 0 & 4 \\ -6 & 8 \end{bmatrix} + 3 \begin{bmatrix} 2 & 0 \\ 0 & -4 \end{bmatrix}$$

$$= \begin{bmatrix} 0 & 4 \\ -6 & 8 \end{bmatrix} + \begin{bmatrix} 6 & 0 \\ 0 & -12 \end{bmatrix}$$

$$2M = \begin{bmatrix} 6 & 4 \\ -6 & -4 \end{bmatrix}$$

$$\therefore M = \begin{bmatrix} 3 & 2 \\ -3 & -2 \end{bmatrix}$$

EXERCISE 6E

Express as a single matrix.

1. $3\begin{bmatrix} 4 \\ 5 \end{bmatrix}$

2. $2\begin{bmatrix} -3 \\ 8 \end{bmatrix}$

3. $3\begin{bmatrix} 3 & 0 \\ 5 & -2 \end{bmatrix}$

4. $4\begin{bmatrix} 2 & 3 \\ 0.5 & -5 \end{bmatrix}$

5. $3\begin{bmatrix} 1 \\ 6 \end{bmatrix} + \begin{bmatrix} 7 \\ 2 \end{bmatrix}$

6. $2\begin{bmatrix} 5 \\ 4 \end{bmatrix} - \begin{bmatrix} 3 \\ -1 \end{bmatrix}$

7. $\frac{1}{2}\begin{bmatrix} 8 & 0 \\ 6 & -2 \end{bmatrix}$

8. $\frac{1}{5}\begin{bmatrix} 10 & 0 \\ 15 & -20 \end{bmatrix}$

9. $4\begin{bmatrix} 5 \\ -1 \end{bmatrix} - 5\begin{bmatrix} 3 \\ -4 \end{bmatrix}$

10. $2\begin{bmatrix} 1 & -2 \\ 3 & 0 \end{bmatrix} - \begin{bmatrix} -5 & 2 \\ 0 & 3 \end{bmatrix}$

11. $A = \begin{bmatrix} 0 & -2 \\ 1 & 0 \end{bmatrix}$ $B = \begin{bmatrix} 0 & -2 \\ 1 & 0 \end{bmatrix}$. Find

   (a) A + 2B

   (b) the matrix C such that A + C = B

12. Find the matrix P which is such that $3P + \begin{bmatrix} 0 & 6 \\ -9 & 2 \end{bmatrix} = 4\begin{bmatrix} 3 & 0 \\ 0 & -4 \end{bmatrix}$

6F MULTIPLICATION OF MATRICES

Matrices may be multiplied only if they are compatible; the number of columns in the first matrix equals to the number of rows in the second matrix.

To perform matrix multiplication follow the following steps:

Step 1 : Write the size of each matrix next to each other.

Step 2 : Circle the middle two numbers; if they are same multiplication is possible.

Step 3 : The remaining two numbers represent the size of the product matrix.

EXAMPLE 1	EXAMPLE 2
Given that $A = \begin{bmatrix} 3 & 2 \\ 1 & 0 \end{bmatrix}$ and $B = \begin{bmatrix} 5 \\ 4 \end{bmatrix}$, express AB as a single matrix. **SOLUTION** $$\begin{bmatrix} 3 & 2 \\ 1 & 0 \end{bmatrix} \times \begin{bmatrix} 5 \\ 4 \end{bmatrix}$$ $$2 \times \boxed{2} \quad \boxed{2} \times 1$$ The circled 2's means that multiplication is possible. The size of the answer must be 2×1. $$\begin{bmatrix} 3 & 2 \\ 1 & 0 \end{bmatrix} \times \begin{bmatrix} 5 \\ 4 \end{bmatrix} = \begin{bmatrix} 3 \times 5 + 2 \times 4 \\ 1 \times 5 + 0 \times 4 \end{bmatrix} = \begin{bmatrix} 23 \\ 5 \end{bmatrix}$$	Given that $A = \begin{bmatrix} 2 & 1 \\ 4 & 3 \end{bmatrix}$ and $B = \begin{bmatrix} 0 & 5 \\ 8 & 7 \end{bmatrix}$, express AB as a single matrix. **SOLUTION** $$\begin{bmatrix} 2 & 1 \\ 4 & 3 \end{bmatrix} \times \begin{bmatrix} 0 & 5 \\ 8 & 7 \end{bmatrix}$$ $$2 \times \boxed{2} \quad \boxed{2} \times 2$$ $$\begin{bmatrix} 2 & 1 \\ 4 & 3 \end{bmatrix} \times \begin{bmatrix} 0 & 5 \\ 8 & 7 \end{bmatrix}$$ $$= \begin{bmatrix} 2 \times 0 + 1 \times 8 & 2 \times 5 + 1 \times 7 \\ 4 \times 0 + 3 \times 8 & 4 \times 5 + 3 \times 7 \end{bmatrix}$$ $$= \begin{bmatrix} 8 & 17 \\ 24 & 41 \end{bmatrix}$$
EXAMPLE 3	**EXAMPLE 4**
Given that $A = \begin{bmatrix} 2 & 1 \\ 4 & 5 \end{bmatrix}$, simplify A^2. **SOLUTION** $$A^2 = A \times A = \begin{bmatrix} 2 & 1 \\ 4 & 5 \end{bmatrix} \times \begin{bmatrix} 2 & 1 \\ 4 & 5 \end{bmatrix}$$ $$= \begin{bmatrix} 2 \times 2 + 1 \times 4 & 2 \times 1 + 1 \times 5 \\ 4 \times 2 + 5 \times 4 & 4 \times 1 + 5 \times 5 \end{bmatrix}$$ $$= \begin{bmatrix} 8 & 7 \\ 28 & 29 \end{bmatrix}$$	Given that $\begin{bmatrix} 3 & 2 & -1 \\ 1 & 0 & 4 \end{bmatrix} \begin{bmatrix} 1 \\ m \\ 3 \end{bmatrix} = \begin{bmatrix} 4 \\ 13n \end{bmatrix}$, find the values of m and n. **SOLUTION** $$\begin{bmatrix} 3 & 2 & -1 \\ 1 & 0 & 4 \end{bmatrix} \begin{bmatrix} 1 \\ m \\ 3 \end{bmatrix} = \begin{bmatrix} 3 \times 1 + 2 \times m - 1 \times 3 \\ 1 \times 1 + 0 \times m + 4 \times 3 \end{bmatrix}$$ $$= \begin{bmatrix} 3 + 2m - 3 \\ 1 + 0m + 12 \end{bmatrix}$$ $$= \begin{bmatrix} 2m \\ 13 \end{bmatrix}$$ $$\begin{bmatrix} 2m \\ 13 \end{bmatrix} = \begin{bmatrix} 4 \\ 13n \end{bmatrix}$$ $\therefore m = 2$ and $n = 1$

EXERCISE 6F

Evaluate the following matrix products

1. $\begin{bmatrix} 5 & 4 \\ 1 & 2 \end{bmatrix} \begin{bmatrix} 3 \\ 2 \end{bmatrix}$	2. $\begin{bmatrix} 6 & 7 \\ 1 & 3 \end{bmatrix} \begin{bmatrix} 1 \\ 5 \end{bmatrix}$
3. $\begin{bmatrix} 5 & 4 \\ 1 & 0 \end{bmatrix} \begin{bmatrix} 3 & 6 \\ 4 & 3 \end{bmatrix}$	4. $\begin{bmatrix} 7 & 2 \\ 4 & 0 \end{bmatrix} \begin{bmatrix} 5 & -2 \\ 1 & 3 \end{bmatrix}$
5. $\begin{bmatrix} 4 & 5 \end{bmatrix} \begin{bmatrix} -3 \\ 2 \end{bmatrix}$	6. $\begin{bmatrix} 4 & 0 & 5 \end{bmatrix} \begin{bmatrix} -3 \\ 2 \\ 1 \end{bmatrix}$
7. $\begin{bmatrix} 3 \\ 6 \end{bmatrix} \begin{bmatrix} 2 & 5 \end{bmatrix}$	8. $\begin{bmatrix} 4 \\ -2 \end{bmatrix} \begin{bmatrix} 7 & 10 \end{bmatrix}$
9. $\begin{bmatrix} 8 & -1 & 1 \\ 4 & 0 & 2 \end{bmatrix} \begin{bmatrix} 2 \\ 5 \\ 3 \end{bmatrix}$	10. $\begin{bmatrix} 3 & 4 \end{bmatrix} \begin{bmatrix} 5 & 0 \\ -1 & 2 \end{bmatrix}$
11. $\begin{bmatrix} 4 & -2 & 5 \end{bmatrix} \begin{bmatrix} 2 & 4 \\ 5 & 0 \\ 3 & -1 \end{bmatrix}$	12. $\begin{bmatrix} 2 & 0 & 3 \end{bmatrix} \begin{bmatrix} 4 & 6 \\ 1 & 3 \\ 5 & 0 \end{bmatrix}$
13. $\begin{bmatrix} 5 & 4 \\ 1 & 0 \end{bmatrix} \begin{bmatrix} 1 & 0 \\ 0 & 1 \end{bmatrix}$	14. $\begin{bmatrix} 6 & 5 \\ 4 & 3 \end{bmatrix} \begin{bmatrix} 1 & 0 \\ 0 & 1 \end{bmatrix}$

15. Given that $\begin{bmatrix} 5 & 2 & -1 \\ 1 & 0 & 4 \end{bmatrix} \begin{bmatrix} 1 \\ m \\ 5 \end{bmatrix} = \begin{bmatrix} 8 \\ 7n \end{bmatrix}$, find the values of m and n.

16. Given that $\begin{bmatrix} 3 & 1 \\ -1 & q \end{bmatrix} \begin{bmatrix} 1 & 2 \\ 0 & 1 \end{bmatrix} = \begin{bmatrix} p & 7 \\ -1 & 4 \end{bmatrix}$, find the value of p and the value of q.

17. $M = \begin{bmatrix} 1 & s \\ r & 6 \end{bmatrix}$ and $N = \begin{bmatrix} 2 & -3 \\ 0 & 8 \end{bmatrix}$

(a) Express 4M – 3N in terms of r and s.

(b) Find N^2

(c) Given that NM = 8M, find the value of r and the value of s.

6G DETERMINING THE POWER OF A MATRIX USING TECHNOLOGY

In this part of the chapter, we are going to make use of technology (CAS) to evaluate the power of matrices where complicated numerical manipulation is required.

EXAMPLES

1. Given that $M = \begin{bmatrix} 1 & 2 & 0 \\ 0 & 5 & 4 \\ -1 & 3 & 6 \end{bmatrix}$, using your CAS to simplify M^2.

    SOLUTION

    $$M^2 = \begin{bmatrix} 1 & 12 & 8 \\ -4 & 37 & 44 \\ -7 & 31 & 48 \end{bmatrix}$$

2. Given that $N = \begin{bmatrix} 1 & 4 & 3 \\ 2 & 0 & 4 \\ 1 & 5 & -3 \end{bmatrix}$, using your CAS to simplify N^3.

    SOLUTION

    $$N^3 = \begin{bmatrix} 60 & 98 & 82 \\ 56 & -6 & 148 \\ 18 & 192 & -116 \end{bmatrix}$$

3. Given that $A = \begin{bmatrix} 3 & 1 & 0 \\ 0 & 6 & 4 \\ -1 & 4 & 3 \end{bmatrix}$ and $B = \begin{bmatrix} 2 & 4 & 3 \\ 5 & 0 & 1 \\ 1 & 8 & -2 \end{bmatrix}$, simplify $A^2 + B^2$.

    SOLUTION

    $$A^2 + B^2 = \begin{bmatrix} 36 & 41 & 8 \\ 7 & 80 & 49 \\ 34 & 23 & 40 \end{bmatrix}$$

EXERCISE 6G

1. Given that $P = \begin{bmatrix} 2 & 3 & 5 \\ 0 & 4 & 1 \\ 1 & 3 & 0 \end{bmatrix}$, simplify P^2.

2. Simplify A^2 if $A = \begin{bmatrix} 1 & 8 & 9 & 2 \\ 1 & 7 & 5 & 1 \\ 2 & 0 & 5 & 0 \\ 3 & 4 & 1 & 7 \end{bmatrix}$

3. Given that $Q = \begin{bmatrix} 2 & -1 & 6 & 3 \\ 1 & 3 & 1 & 1 \\ 4 & 1 & 4 & 2 \\ 3 & 2 & 3 & 4 \end{bmatrix}$, simplify Q^2.

4. Simplify $A^2 + B^2$, where $A = \begin{bmatrix} 4 & 2 & 1 \\ 1 & 3 & 4 \\ 5 & 1 & 0 \end{bmatrix}$ and $B = \begin{bmatrix} 0 & 2 & 1 \\ 1 & 0 & 0 \\ 1 & 1 & 0 \end{bmatrix}$

6H ONE STAGE ADJACENCY MATRIX

An **adjacency matrix** is a simple way of representing which vertices, also named nodes, of a diagram are adjacent to which other vertices.

Consider the diagram below which shows the different routes that exist between 4 suburbs.

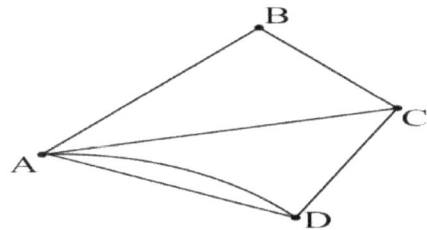

To figure the adjacency matrix, we have to find out the different direct ways of travelling from one suburb to the other.

From suburb A there is 1 direct route to suburb B, 1 to C and 2 to D.
From suburb B there is 1 direct route to A, 1 to C and none to D.
From suburb C there is 1 direct route to A, 1 to B and 1 to D.
From suburb D there are 2 direct routes to A, none to B and 1 to C.

Hence the adjacency matrix is

		TO			
		A	B	C	D
FROM	A	0	1	1	2
	B	1	0	1	0
	C	1	1	0	1
	D	2	0	1	0

EXAMPLE
Write down the adjacency matrix for the network below showing roads existing between three towns.

SOLUTION
Note that the road from A to B is a one-way traffic.
There is no direct route from B to A.
Similarly there are 2 direct route from B to C but only 1 from C to B.
Hence the adjacency matrix is as shown.

		FROM		
		A	B	C
TO	A	0	1	1
	B	0	0	2
	C	1	1	0

EXERCISE 6H

Write down the adjacency matrix for each of the following network.

Network	Adjacency matrix
1.	
2.	
3.	
4.	
5.	

6I TWO STAGE ADJACENCY MATRIX

A two stage adjacency matrix shows the number of two-length path from one node to the others.

Consider the network on the right showing the relationship between 3 friends A, B and C.
A considers B to be a friend but B doesn't feel the same. Likewise, B considers C to be a friend while C feels otherwise.

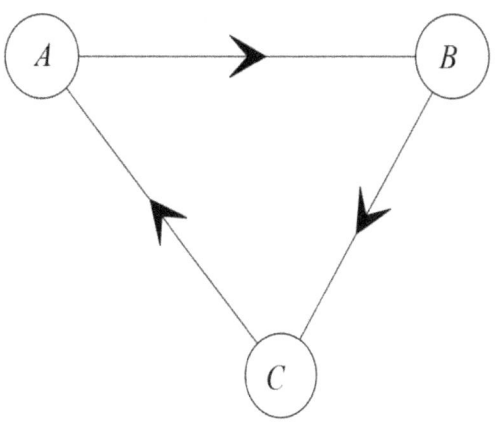

An adjacency matrix for the given network is

		FROM		
		A	B	C
TO	A	0	1	0
	B	0	0	1
	C	1	0	0

Consider a two stage matrix.
There is 1 path from A to C ($A \to B \to C$), 0 from A to B and 0 from A to itself.
Similarly, there is 1 path from B to A ($B \to C \to A$), 0 from B to C and 0 from B to itself.
And there is 1 path from C to B ($C \to A \to B$), 0 from C to A and 0 from C to itself.
Hence a two stage adjacency matrix is given as under:

		FROM		
		A	B	C
TO	A	0	0	1
	B	1	0	0
	C	0	1	0

Let $M = \begin{bmatrix} 0 & 1 & 0 \\ 0 & 0 & 1 \\ 1 & 0 & 0 \end{bmatrix}$ be the adjacency matrix.

Now $M^2 = \begin{bmatrix} 0 & 0 & 1 \\ 1 & 0 & 0 \\ 0 & 1 & 0 \end{bmatrix}$ and this is equivalent the two stage adjacency matrix.

In the examples that follow, we are going to investigate if squaring the one stage adjacency matrix is equivalent to the two stage adjacency matrix.

EXAMPLE

For the network,

    (a) Write down the adjacency matrix M,
    (b) Write down the two stage adjacency matrix N,
    (c) Calculate the square of the adjacency matrix M^2.
    (d) Compare your answers to (b) and (c).

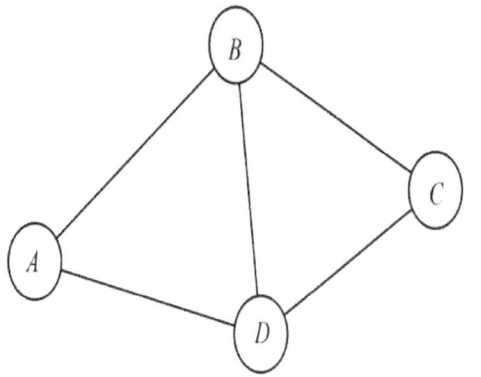

SOLUTION

(a) Adjacency matrix M is

		FROM			
		A	B	C	D
TO	A	0	1	0	1
	B	1	0	1	1
	C	0	1	0	1
	D	1	1	1	0

Or $M = \begin{bmatrix} 0 & 1 & 0 & 1 \\ 1 & 0 & 1 & 1 \\ 0 & 1 & 0 & 1 \\ 1 & 1 & 1 & 0 \end{bmatrix}$

(b) Two stage adjacency matrix is

		FROM			
		A	B	C	D
TO	A	2	1	2	1
	B	1	3	1	2
	C	2	1	2	1
	D	1	2	1	3

There are 2 paths from A to itself $(A \to B \to A$ & $A \to D \to A)$, *1 path from A to B* $(A \to D \to B)$, *2 paths from A to C* $(A \to B \to C$ & $A \to D \to C)$ *and 1 path from A to D* $(A \to B \to D)$.

Or $N = \begin{bmatrix} 2 & 1 & 2 & 1 \\ 1 & 3 & 1 & 2 \\ 2 & 1 & 2 & 1 \\ 1 & 2 & 1 & 3 \end{bmatrix}$

(c) $M^2 = \begin{bmatrix} 2 & 1 & 2 & 1 \\ 1 & 3 & 1 & 2 \\ 2 & 1 & 2 & 1 \\ 1 & 2 & 1 & 3 \end{bmatrix}$

(d) Clearly $M^2 = N$, implying that a two stage matrix is equivalent to the square of a one stage adjacency matrix.

CHAPTER 6 : MATRICES 95

EXERCISE 6I

For each of the following determine the one stage adjacency matrix and the two stage adjacency matrix. Investigate using your CAS calculator whether the square of the one stage adjacency matrix is equivalent to the two stage adjacency matrix.

	One stage adjacency matrix	Two stage adjacency matrix
1.		
2.		
3.		
4.		

6J SOCIAL INTERACTION AS A MATRIX

Matrix as we have seen so far has various uses and importance. In this section, we are going to show that matrix can be used to portray social interaction within a social group.

The diagram on the right shows a group of four students meeting at an orientation day at UWA. All the four students have enrolled in the same course. Student A happens to have met all the other 3 student's parents. Student B, on the other hand, knew only student C's parents prior to this meeting. Student C had already met the parents of student A, B and D.

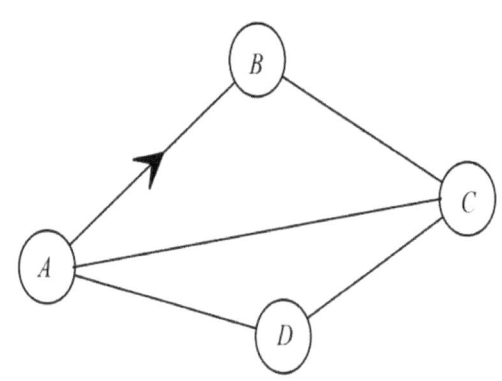

Let us assume the number 0 indicates someone did not meet the other's parent before and the number 1 indicating they have already met their parents. The information can then be shown as a matrix as under:

		FROM			
		A	B	C	D
TO	A	0	1	1	1
	B	0	0	1	0
	C	1	1	0	1
	D	1	0	1	0

OR $S = \begin{bmatrix} 0 & 1 & 1 & 1 \\ 0 & 0 & 1 & 0 \\ 1 & 1 & 0 & 1 \\ 1 & 0 & 1 & 0 \end{bmatrix}$

It is quite obvious that someone should have met his or her own parent at one point. This is the reason we have zeroes along the leading diagonal.

EXERCISE 6J

1. Five middle school students did a survey asking in discretion "Do you like me?" The result is shown in the diagram on the right. Use 1 to represent someone liking the other and 0 for disliking, complete the matrix below.

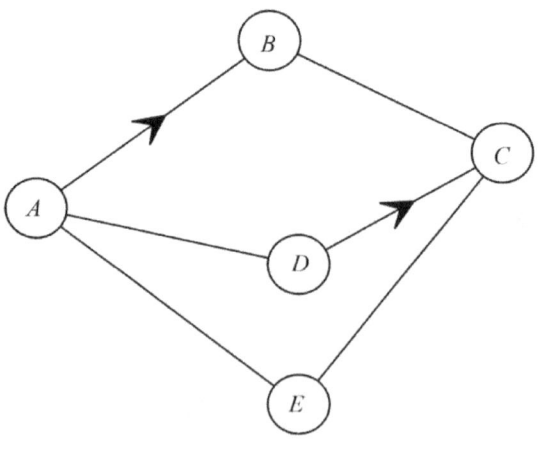

	A	B	C	D	E
A	0				
B		0			
C			0		
D				0	
E					0

2. A year 12 IT teacher did a survey from his top 6 students to find out who emailed each other before. An analysis of the responses produced the diagram as shown.

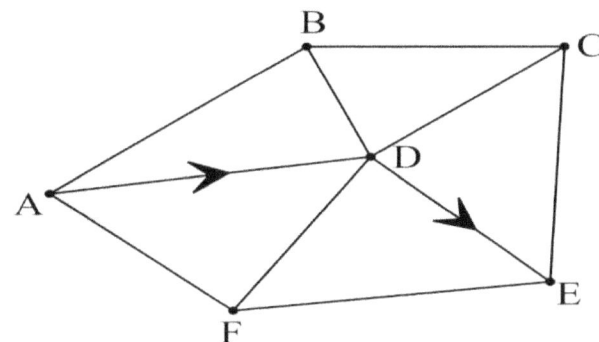

Complete the matrix below using

- 0 to show someone has emailed the other.
- 1 to show someone hasn't emailed the other
- 0 for someone emailing himself or herself.

	A	B	C	D	E	F
A	0					
B		0				
C			0			
D				0		
E					0	
F						0

3. The diagram on the right shows "who has been on camp with whom in a particular year level" among a group of five classmates Alex, Bob, Carl, David and Ethan.
Complete the matrix below using

- 0 to show a pair has not been on camp together,
- 1 to show a pair has been on camp together,
- 0 for someone being on camp accompanying himself or herself.

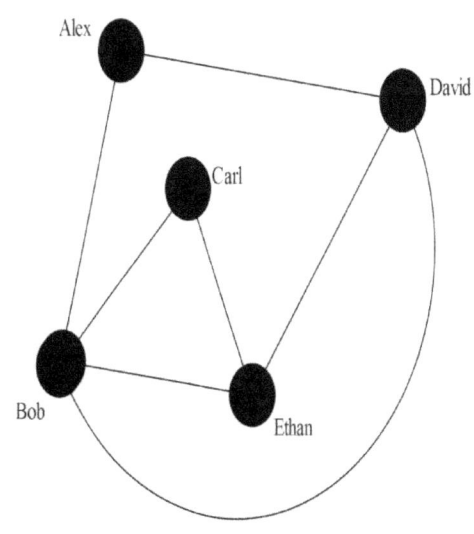

	Alex	Bob	Carl	David	Ethan
Alex	0				
Bob		0			
Carl			0		
David				0	
Ethan					0

4. "Do you have my mobile number?" The diagram shows the result of the above question asked to six friends who all work together at a fast food place.

Complete the matrix below using
- 0 showing not having mobile number,
- 1 showing having mobile number,
- 0 for having someone's own mobile number.

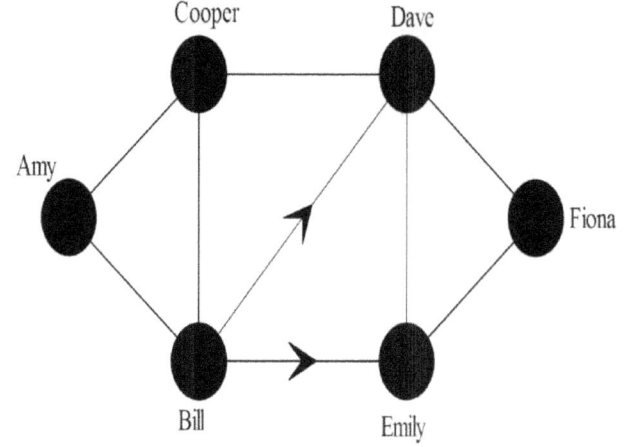

	Amy	Bill	Cooper	Dave	Emily	Fiona
Amy	0					
Bill		0				
Cooper			0			
Dave				0		
Emily					0	
Fiona						0

5. "Have you slept over at my house?" Five friends' response to the above question is shown in the diagram on the right. Complete the matrix below using
- 0 showing not slept over,
- 1 showing having slept over,
- 0 for having slept at own house.

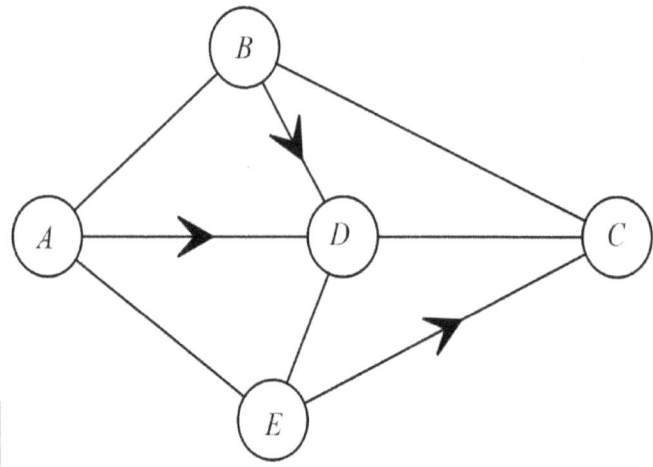

	A	B	C	D	E
A	0				
B		0			
C			0		
D				0	
E					0

CHAPTER 6 : MATRICES

6K STORING, DISPLAYING INFORMATION, MODEL AND SOLVE PROBLEMS

EXAMPLE 1

The table shows the number of games played and the results of four teams in a football league.

	Played	Won	Drawn	Lost
City	8	4	1	3
United	7	3	0	4
Tigers	8	4	0	4
Wolf	7	2	1	4

A win earns 3 points, a draw 1 point and a loss 0 point. Write down two matrices which on multiplication display in their product the total number of points earned by each team and hence calculate these totals.

Solution

$$\begin{bmatrix} 4 & 1 & 3 \\ 3 & 0 & 4 \\ 4 & 0 & 4 \\ 2 & 1 & 4 \end{bmatrix} \begin{bmatrix} 3 \\ 1 \\ 0 \end{bmatrix} = \begin{bmatrix} 13 \\ 9 \\ 12 \\ 7 \end{bmatrix} \quad \text{or} \quad \begin{bmatrix} 3 & 1 & 0 \end{bmatrix} \begin{bmatrix} 4 & 3 & 4 & 2 \\ 1 & 0 & 0 & 1 \\ 3 & 4 & 4 & 4 \end{bmatrix} = \begin{bmatrix} 13 & 9 & 12 & 7 \end{bmatrix}$$

EXAMPLE 2

A circus show is held over a three-day period – Friday, Saturday and Sunday. The table below shows the entry price per day for an adult and for a child, and the number of adults and children attending on each day.

	Friday	Saturday	Sunday
Price ($) - Adult	12	10	10
Price ($) – Child	6	5	5
Number of adults	360	190	400
Number of children	50	60	150

(i) Write down two matrices such that their product will give the amount of entry money paid on Friday and hence calculate this product.

$$\begin{bmatrix} 360 & 50 \end{bmatrix} \begin{bmatrix} 12 \\ 6 \end{bmatrix} = [4620] \quad \text{or} \quad \begin{bmatrix} 12 & 6 \end{bmatrix} \begin{bmatrix} 360 \\ 50 \end{bmatrix} = [4620]$$

(ii) Write down two matrices such that their product will give the amount of entry money paid on Sunday and hence calculate this product.

$$\begin{bmatrix} 400 & 150 \end{bmatrix} \begin{bmatrix} 10 \\ 5 \end{bmatrix} = [4750]$$

(iii) Calculate the percentage increase in revenue on Sunday compared to Friday.

$$\frac{4750 - 4620}{4620} \times 100 = 2.81\%$$

MATHEMATICS APPLICATIONS UNIT 1

EXERCISE 6K

1.

Place Team	1st	2nd	3rd
Hobbits	4	5	3
Dwarfs	6	0	6
Hogwarts	2	7	3

The table shows the results achieved by three teams in twelve events of an athletics competition. In each event, 1st place scores 5 points, 2nd place scores 3 points, and 3rd place scores 1 point.

(i) Write down two matrices whose product shows the total number of points scored by each team.

(ii) Evaluate this product of matrices.

2. During a numeracy test, students take three multiple-choice tests, each with ten questions. A correct answer earns 5 marks. If no answer is given 1 mark is scored. An incorrect answer loses 2 marks. A student's final total mark is the sum of 20% of the mark in test 1, 30% of the mark in test 2 and 50% of the mark in test 3. One student's responses are summarized in the table below.

	Test 1	Test 2	Test 3
Correct answer	7	6	5
No answer	2	0	1
Incorrect answer	1	4	4

Write down three matrices such that matrix multiplication will give this student's final total mark and hence find this total mark.

3. Western Air has a fleet of aircraft consisting of 3 aircraft of type P, 5 of type Q, 2 of type R and 10 of type S. The aircraft have 3 classes of seat known as Economy, Business and First. The table below shows the number of these seats in each of the 4 types of aircraft.

	Economy	Business	First
P	300	30	10
Q	180	40	15
R	140	20	5
S	110	5	0

(i) Write down two matrices whose product shows the total number of seats in each class.

(ii) Evaluate this product of matrices.

On a particular day, each aircraft made one flight. 5% of the Economy seats were empty, 10% of the Business seats were empty and 20% of the First seats were empty.

(iii) Write down a matrix whose product with the matrix found in part (ii) will give the total number of empty seats on that day.

(iv) Evaluate this total.

4. Great Auto sells three models of racing cars, A, B and C. The table below shows the numbers of each model sold over a four-week period and the cost of each model in $.

Week \ Model	A	B	C
1	8	12	4
2	7	10	2
3	10	12	0
4	6	8	4
Costs($)	250	400	700

In the first two weeks the shop banked 30% of all money received, but in the last two weeks the shop only banked 20% of all money received.

GREAT AUTO CO.

(i) Write down three matrices such that matrix multiplication will give the total amount of money banked over the four-week period.

(ii) Hence evaluate this total amount.

CHAPTER 7

THE THEOREM OF PYTHAGORAS

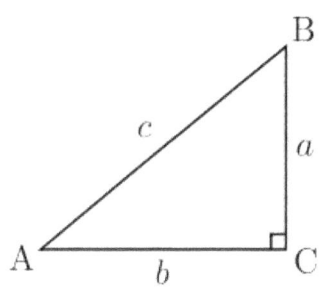

A triangle that contains a 90° angle is called a right –angled triangle as shown.
The longest side of the triangle is called the **hypotenuse**.

The Pythagoras theorem states

> *For any right-angled triangle, the square of the length of the hypotenuse is equal to the sum of the squares of the lengths of the two shorter sides.*

For the triangle on the right

$AB^2 = AC^2 + BC^2$ or $c^2 = a^2 + b^2$

7A HOW TO PROVE A TRIANGLE IS RIGHT ANGLED?

To prove a triangle is right-angled, we have to use the converse of Pythagoras theorem. In other words, the sum of the squares of the lengths of the two shorter sides must be equal the square of the length of the hypotenuse.

EXAMPLE 1

Show that the triangle is right-angled.

SOLUTION

Let $c = 10, a = 6$ and $b = 8$

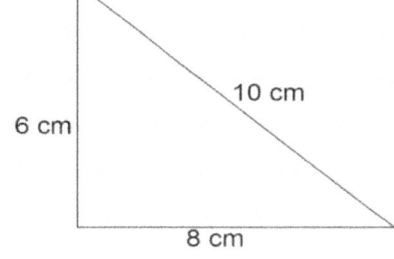

$6^2 + 8^2 = 36 + 64 = 100$

And $10^2 = 100$.

As $c^2 = a^2 + b^2$ holds for the given triangle, it is proven that it is a right-angled triangle.

EXAMPLE 2

Show that the triangle is **NOT** right-angled.

SOLUTION

Let $c = 4, a = 2$ and $b = 3$

$2^2 + 3^2 = 4 + 9 = 13$

And $4^2 = 16$.

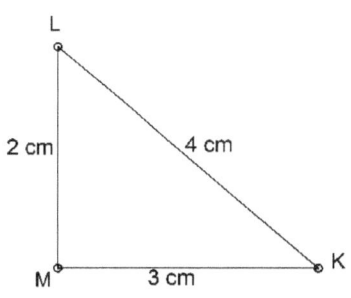

As $c^2 = a^2 + b^2$ does not hold for the given triangle, it is proven that it is not a right-angled triangle.

EXERCISE 7A

Which of the following triangles are right-angled?

1.

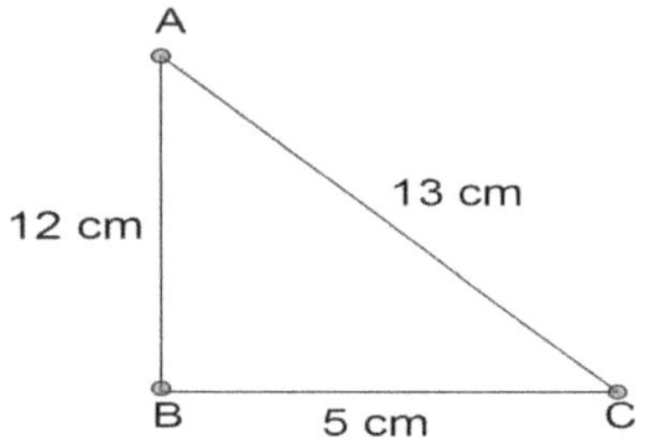

2.

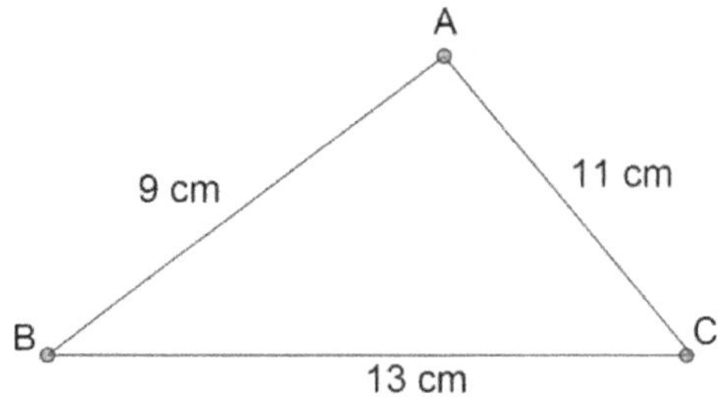

3.

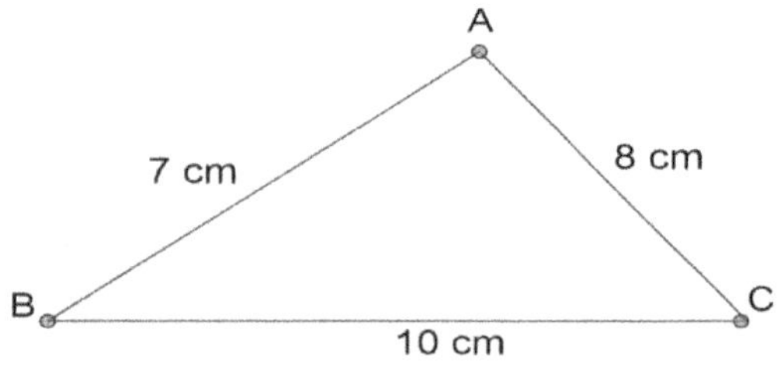

4.

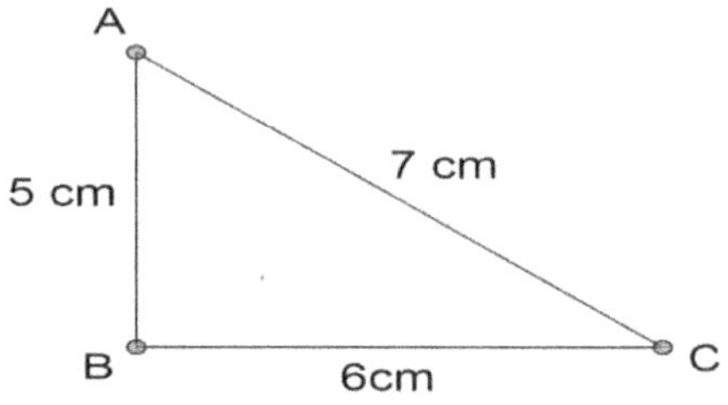

5.

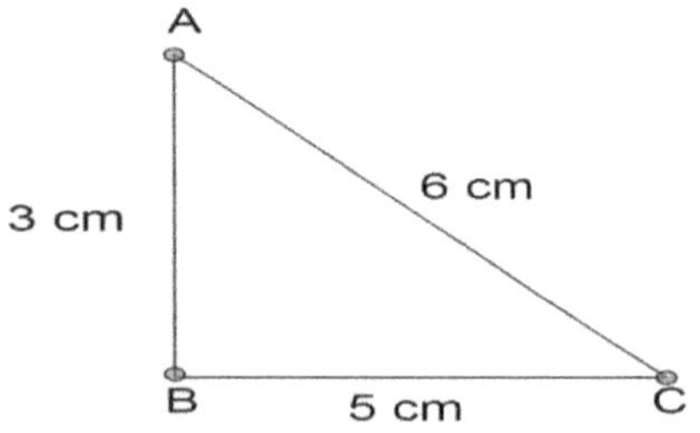

6.

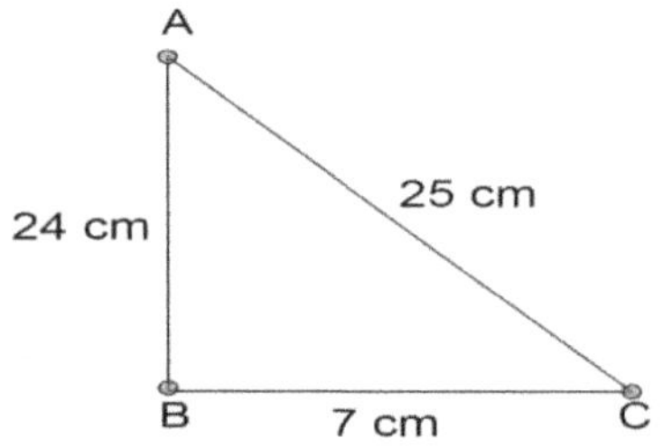

CHAPTER 7 : THE THEOREM OF PYTHAGORAS

7B MENTAL SKILLS

CLASS ACTIVITY 1

Evaluate the following without the use of a calculator.

1. 5^2	2. 7^2	3. 9^2
4. 10^2	5. 30^2	6. 50^2
7. 200^2	8. 600^2	9. $\left(\dfrac{4}{5}\right)^2$
10. $\sqrt{9}$	11. $\sqrt{64}$	12. $\sqrt{121}$
13. $\sqrt{169}$	14. $\sqrt{225}$	15. $\sqrt{4900}$
16. $\sqrt{9+16}$	17. $\sqrt{40-36}$	18. $\sqrt{36+64}$
19. $\sqrt{144+25}$	20. $\sqrt{600-200}$	21. $\sqrt{1200+400}$

Evaluate the following to 2 decimal places, using your calculator where necessary.

22. $\sqrt{5^2+4^2}$	23. $\sqrt{7^2+11^2}$	24. $\sqrt{25^2+16^2}$
25. $\sqrt{15^2+20^2}$	26. $\sqrt{17^2+18^2}$	27. $\sqrt{2^2+3^2}$

CLASS ACTIVITY 2

For each of the following write down the relationship between the pronumerals. The first two has been done as examples.

1.

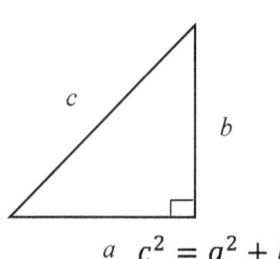

$c^2 = a^2 + b^2$

2.

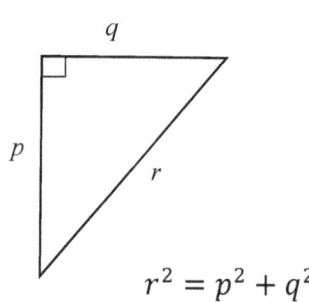

$r^2 = p^2 + q^2$

3.

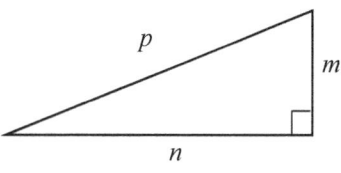

4.

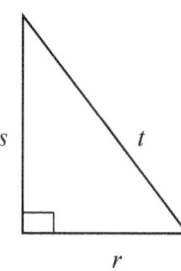

5.

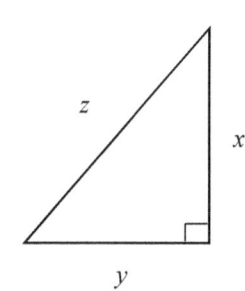

6.

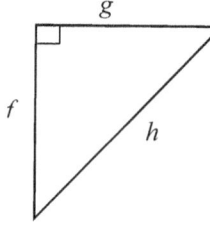

For the right-angled triangles shown below, circle the true statement.

7.

A. $q^2 = p^2 + 11^2$

B. $11^2 = p^2 + q^2$

C. $p^2 = q^2 + 11^2$

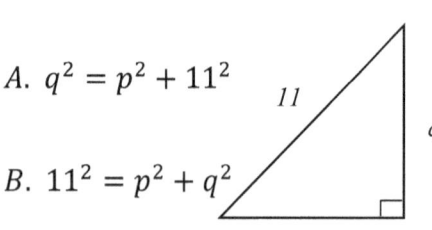

8.

A. $y^2 = x^2 + 4^2$

B. $4^2 = x^2 + y^2$

C. $x^2 = y^2 + 4^2$

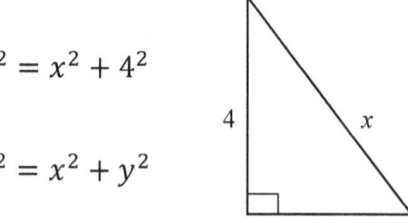

7C FINDING THE LENGTH OF THE HYPOTENUSE

If two shorter sides (also known as legs) of a right-angled are known, we can use Pythagoras' Theorem to find the length of the hypotenuse.

EXAMP1E 1	EXAMPLE 2
Use Pythagoras' theorem to find the length of the unknown side. 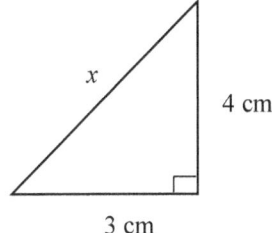 **SOLUTION** Using $c^2 = a^2 + b^2$ Here $a = 3, b = 4, c = x$ Substituting the values, we have $x^2 = 3^2 + 4^2 = 9 + 16 = 25$ $\therefore x = \sqrt{25} = 5 \ (ignore \ x = -5)$	Use Pythagoras' theorem to find the length of y, giving your answer to 2 decimal places. 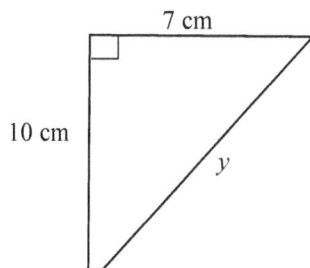 **SOLUTION** Using $c^2 = a^2 + b^2$ Here $a = 7, b = 10, c = y$ Substituting the values, we have $y^2 = 7^2 + 10^2 = 49 + 100 = 149$ $\therefore y = \sqrt{149} = 12.21$

EXAMPLE 3

The diagram shows a trapezium subdivided into a square of side 12 cm and a triangle. The length of the longer parallel side is 17 cm. Find the value of x.

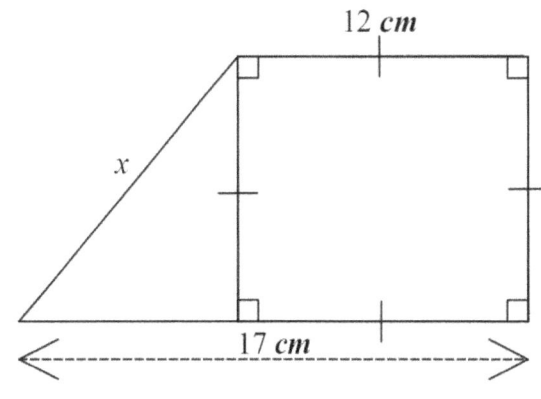

SOLUTION

The two shorter sides of the triangle are 12 cm and 17 − 12 = 5 cm.

Using $c^2 = a^2 + b^2$

Here $a = 5, b = 12, c = x$

Substituting the values, we have

$x^2 = 5^2 + 12^2 = 25 + 144 = 169$

$\therefore x = \sqrt{169} = 13 \ cm.$

EXERCISE 7C

Use Pythagoras' theorem to find the length of the unknown side.

1. Triangle with legs 8 cm and 6 cm, hypotenuse x.

2. Triangle with legs 12 cm and 9 cm, hypotenuse y.

3. Triangle with legs 24 cm and 7 cm, hypotenuse z.

4. Triangle with legs 9 cm and 8 cm, hypotenuse x.

5. Rectangle with sides 8 cm and 15 cm, diagonal x.

6. Triangle with a vertical height of 24 cm meeting the base, dividing it into 7 cm and 10 cm segments; x is the left side, y is the right side.

CHAPTER 7 : THE THEOREM OF PYTHAGORAS

7D FINDING THE LENGTH OF THE SHORTER SIDES

The length of one of the shorter sides of a right-angled triangle can be found if we know the length of the hypotenuse and one of the shorter sides.

EXAMP1E 1	EXAMPLE 2
Use Pythagoras' theorem to find the length of the unknown side. 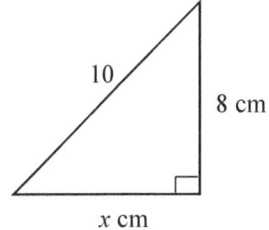	Use Pythagoras' theorem to find the length of x, giving your answer to 2 decimal places. 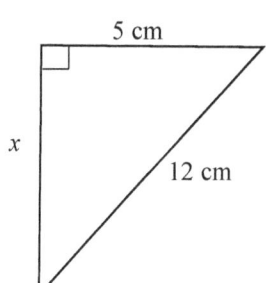
SOLUTION Using $c^2 = a^2 + b^2$ Here $a = x, b = 8, c = 10$ Substituting the values , we have $$10^2 = x^2 + 8^2$$ $$x^2 = 10^2 - 8^2$$ $$= 100 - 64 = 36$$ $$\therefore x = \sqrt{36} = 6 \ (ignore \ x = -6)$$	**SOLUTION** Using $c^2 = a^2 + b^2$ Here $a = x, b = 5, c = 12$ Substituting the values , we have $$12^2 = x^2 + 5^2$$ $$x^2 = 12^2 - 5^2$$ $$= 144 - 25 = 119$$ $$\therefore x = \sqrt{119} = 10.91$$

EXAMPLE 3

The diagram shows a trapezium subdivided into a square of side 10 cm and a triangle. Find the value of x.

SOLUTION

Using $c^2 = a^2 + b^2$

Here $a = x, b = 10, c = 14$

Substituting the values , we have

$$14^2 = x^2 + 10^2 \quad \therefore x^2 = 14^2 - 10^2$$

$$= 196 - 100 = 96$$

$$\therefore x = \sqrt{96} = 9.80 \ cm$$

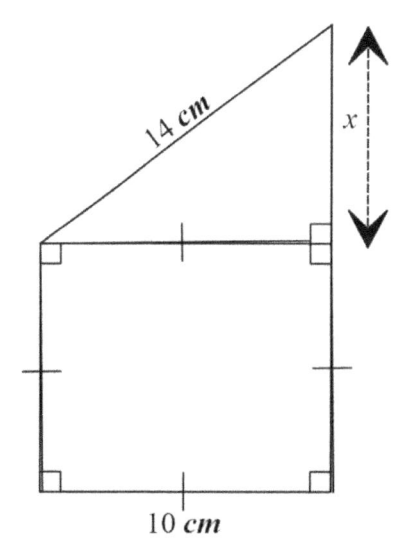

EXERCISE 7D

Use Pythagoras' theorem to find the length of the unknown side.

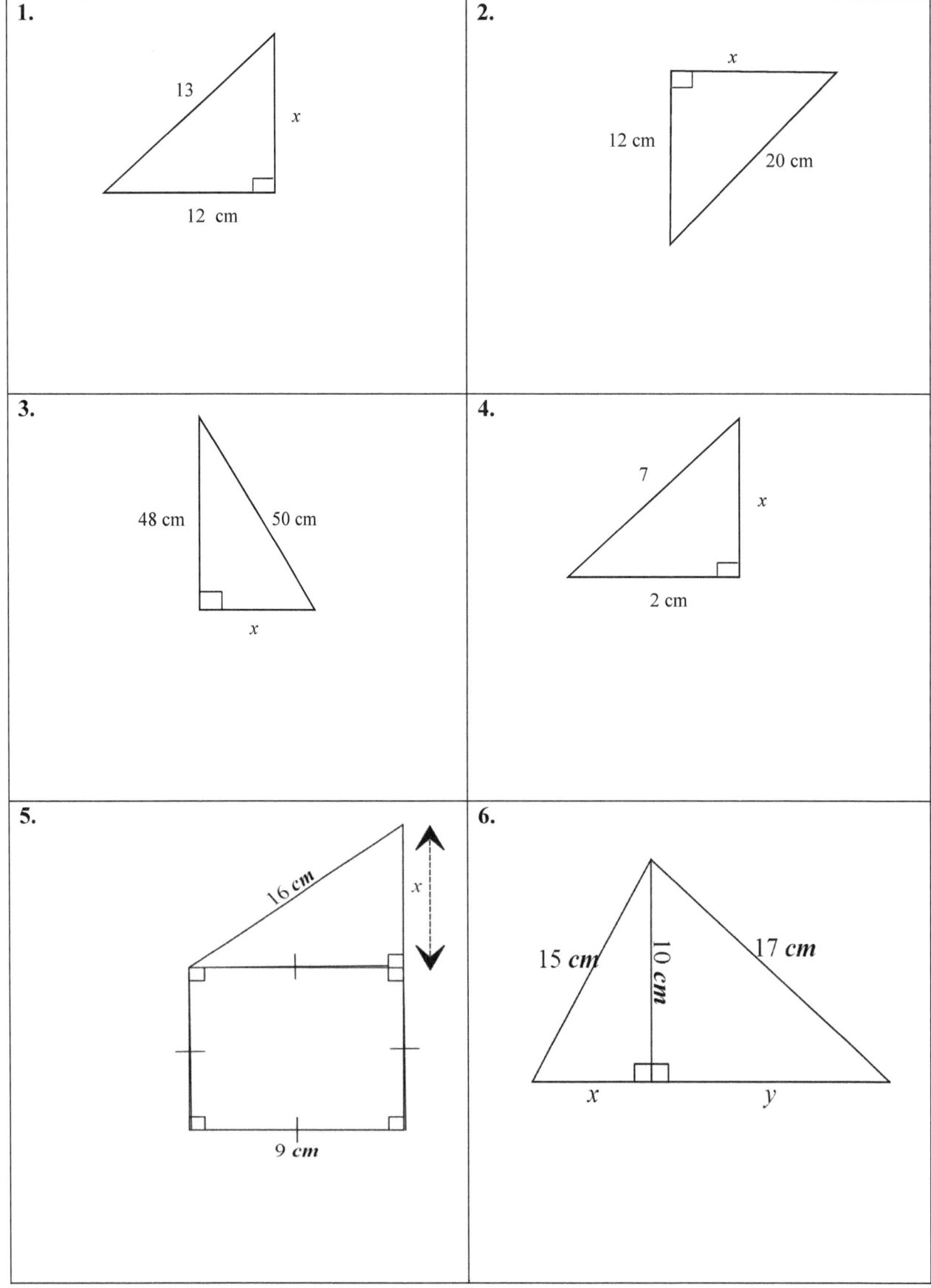

7E   APPLICATIONS IN TWO DIMENSIONS

It is true that one can use Pythagoras theorem in maths problems. But there are other uses too. It is widely used in the world of architecture and in construction of buildings. It is also used to find the speed of sound in water. Furthermore, geologists make use of Pythagoras Theorem to determine the centre of an earthquake.

To solve an application problem with Pythagoras Theorem, follow these simple steps:

- Draw a right-angled triangle to represent the situation
- Identify the right angle, legs and the hypotenuse of the triangle
- Use variables such as x or y to label the unknown sides
- Use Pythagoras Theorem to determine the length of the unknown side.

EXAMPLE 1

A 10 m ladder leans against a wall. The foot of the ladder is 5 m from the base of a vertical wall. How high up the wall does the ladder reach?

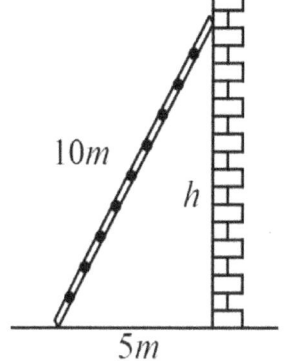

SOLUTION

Using $c^2 = a^2 + b^2$

Here $a = h, b = 5, c = 10$

Substituting the values, we have

$$10^2 = h^2 + 5^2$$

$$x^2 = 10^2 - 5^2 = 100 - 25 = 75$$

$\therefore x = \sqrt{75} = 8.66\ m$

EXAMPLE 2

Sarah is practising her cycling session in regards to the upcoming Tour de France. On a Saturday morning, starting from home she cycled 300 km West and then 100 km South.

(a)   Draw a well-labelled diagram to illustrate the situation.

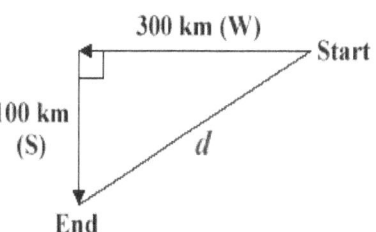

(b)   How far is she from home after the 400 km tour on bike?

Using $c^2 = a^2 + b^2$

Here $a = 3, b = 4, c = d$

$$d^2 = 100^2 + 300^2 = 100\ 000$$

$\therefore d = \sqrt{100000} = 316.23\ km$

EXERCISE 7E

1. A ladder of length 13 m is leaning against a vertical wall. The ladder reaches 12 m up the wall. How far is the foot of the ladder from the bottom of the wall on horizontal ground?

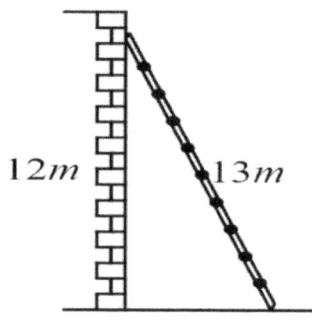

2. Johnny is rapelling (descending at the end of a rope) at Bayside park. His task is to slide down a rope starting 20 m high and goes on for a horizontal distance of 50 m across the river.

   If Johnny did the rapelling 4 times, calculate the total distance he descended along the rope.

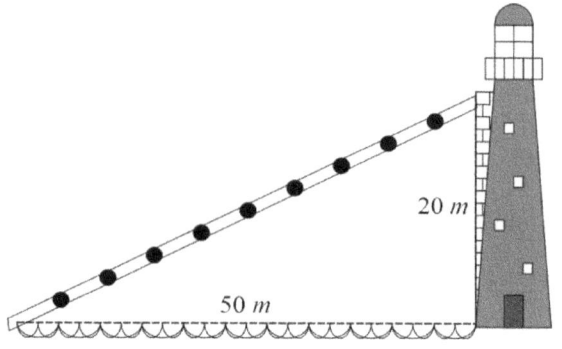

3. Find the length of the diagonal of a rectangular swimming pool whose dimensions are 36 m by 18 m.

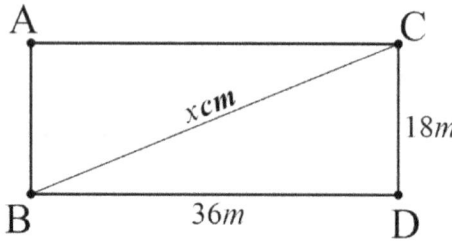

4. As part of his training for the Formula Two competition coming soon, Alex does 50 round trips from A to B. Calculate the total distance travelled, giving your answer in km.

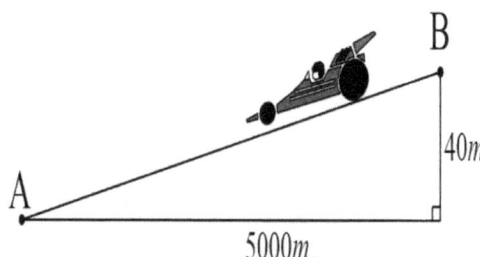

5. In the diagram, D is the point on CA produced such that BD = 14.3 m.
   AB = 10.2 m and AC = 6 m.
   Find the length of AD.

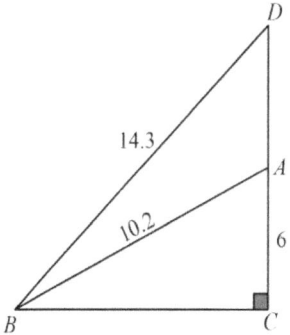

6. PQ is a chord of a circle, centre O.
   X is the midpoint of PQ.
   OX = 6 cm and the radius of the circle is 10 cm.
   Calculate the length of PQ.

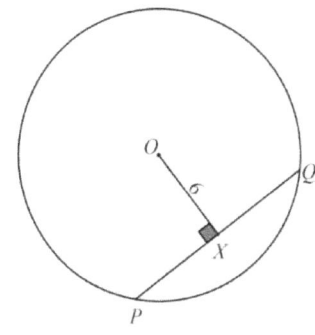

7. Alan bought a kite for his son's 12th birthday. The measurements are shown in the diagram. The outside frame is made of Jarrah. Calculate the length of wood used.

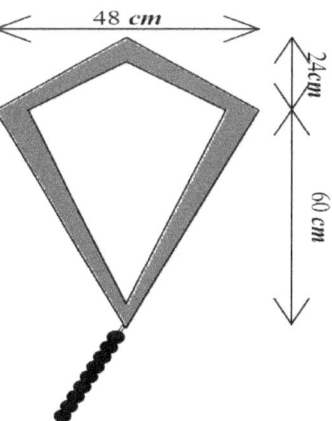

8. A vertical flagpole stands on a horizontal wooden base. The flagpole is held by two iron rods of lengths 16m and 22m as shown in the diagram. Use the measurements to calculate the length of LM.

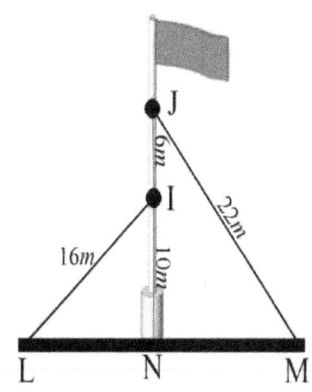

7F APPLICATIONS IN THREE DIMENSIONS

EXAMPLES

1. The diagram shows a cylinder of height 12cm and having radius 4 cm. Find the length of the longest rod that would fit in the cylinder.

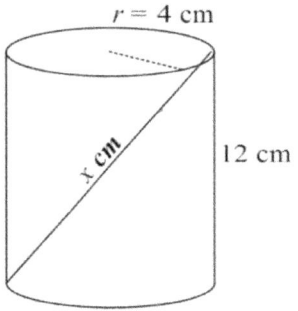

SOLUTION

The radius being 4 cm, the diameter will be 8 cm as shown.

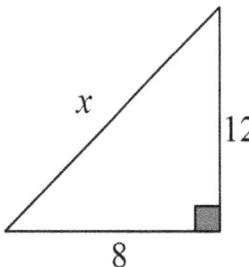

$$\therefore x = \sqrt{12^2 + 8^2} = 14.42 \text{ cm}$$

2. A right circular cone has base radius 8 m and perpendicular height 15 m. Find the slant height.

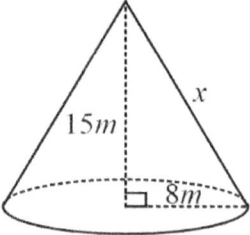

SOLUTION

$$x = \sqrt{8^2 + 15^2} = 17 \text{ m}$$

EXERCISE 7F

1. The diagram shows a rectangular prism. Work out the length of AB.

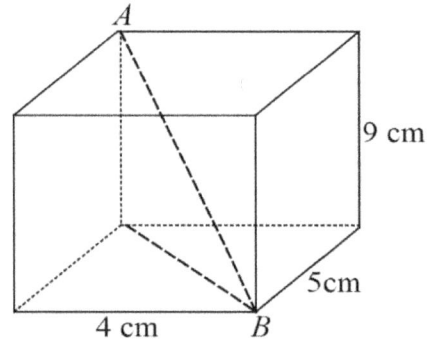

2. The diagram shows a cylinder of height 10cm and having radius 3.6 cm. Find the length of the longest rod that would fit in the cylinder.

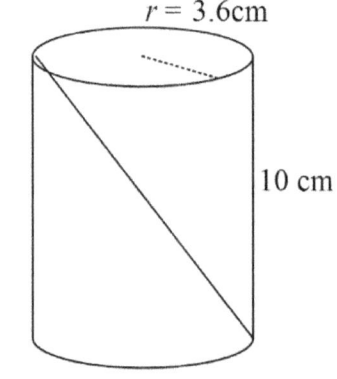

3. Use Pythagoras Theorem to calculate the length of

   (a) QR

   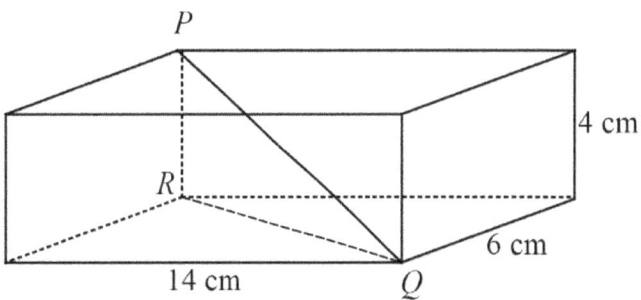

   (b) PQ

4. Thomas just had a shed installed in his backyard. He is planning to lay Christmas lights along the edge AB. Calculate the length of lights thread he will need.

   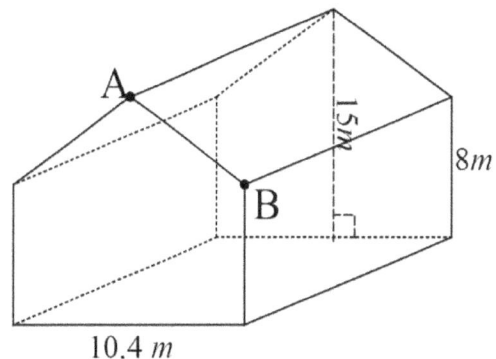

5. Use Pythagoras Theorem to calculate

   (a) CE,

   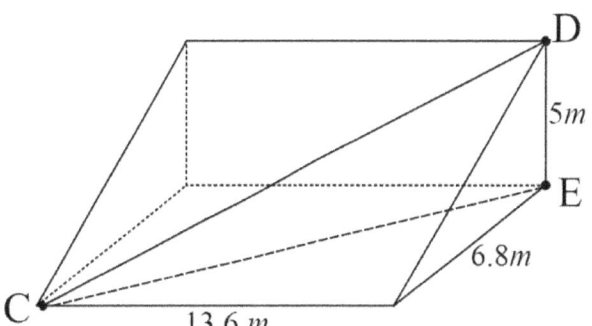

   (b) CD.

6. Peter's Christmas present box measures 4 cm long, 3 cm wide and 12 cm long. Peter was investigating the longest piece of candy that could fit in his box. He was able to figure out that the candy must start at A and end at B as shown.
Use Pythagoras Theorem to find the values of x and y.

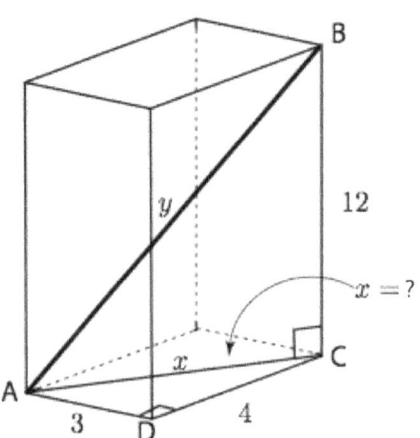

7. A spider crawls along a wooden stick placed inside a rectangular prism of length 4 m, width 2 m and height 4 m as shown. How far does the spider crawl if it goes from one end to the other end of the stick.

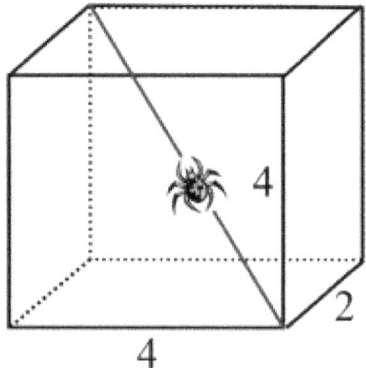

8. The base of a pyramid is a square with diagonals of length 12 cm.
The sloping faces are isosceles triangles with equal sides of length 10 cm.
The height of the pyramid is h cm.
Calculate h.

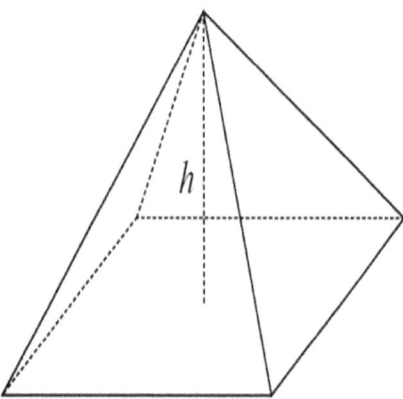

CHAPTER 8

PERIMETER AND AREA

8A PERIMETER

Perimeter is the total distance around the outside of a two-dimensional (2D) shape. To find the perimeter of a shape, we need to add the length of all the sides, including curved sides in some cases. Perimeter can be measured in different units such as metres, inches, km, feet etc..

EXERCISE 8A

Calculate the perimeters of these shapes:

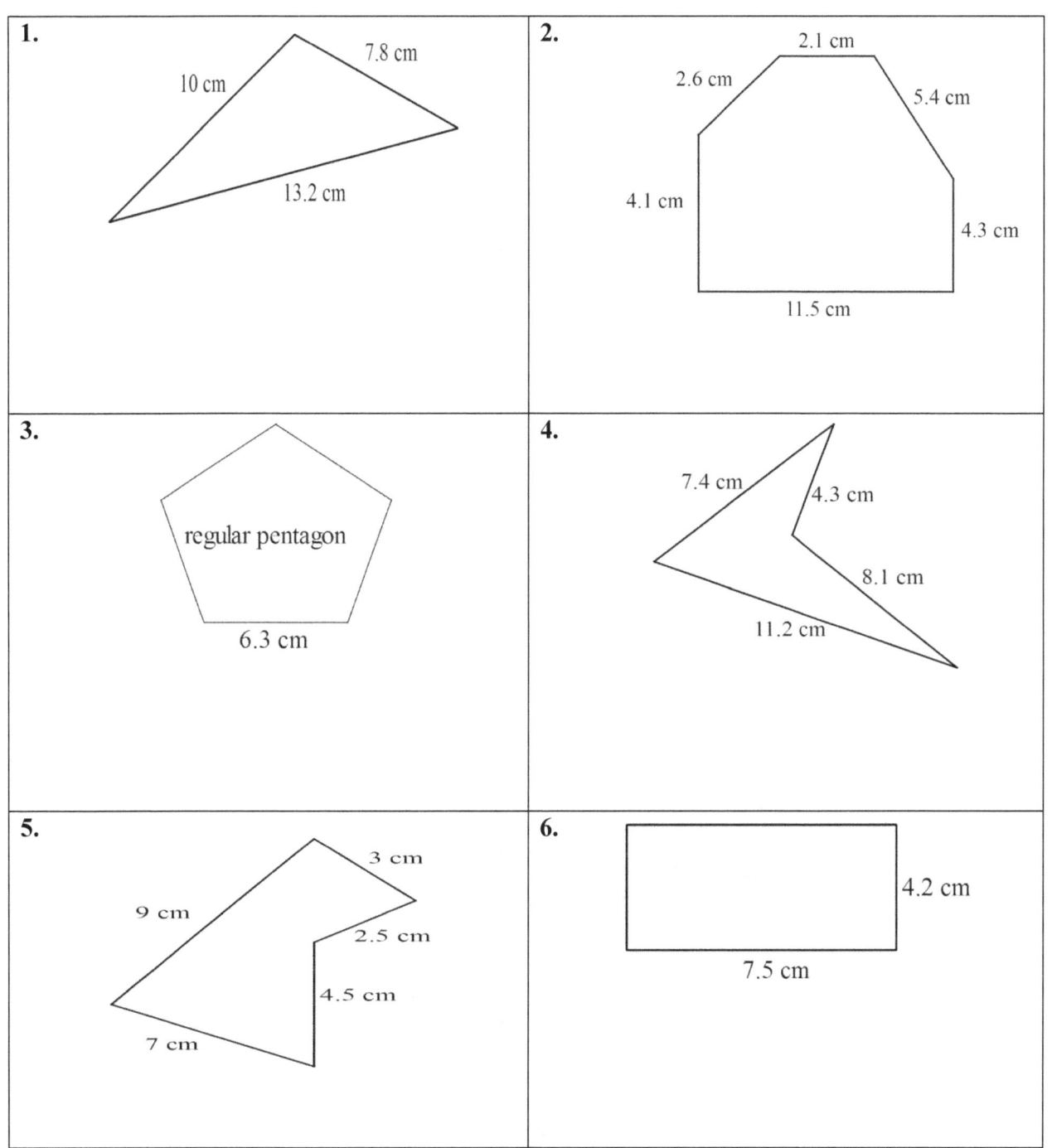

7. The diagram shows a rectangle and a square.
   If they have equal perimeters, what is the length of one side of the square?

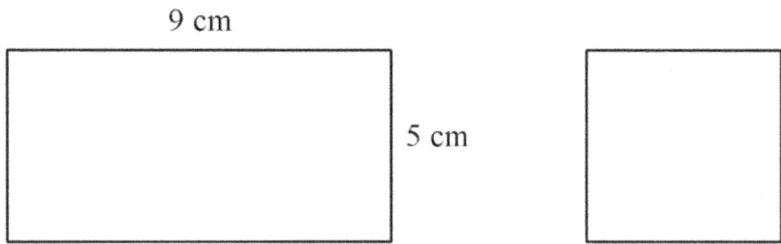

In these composite shapes all the lengths are in centimetres. Work out the perimeter.

8.

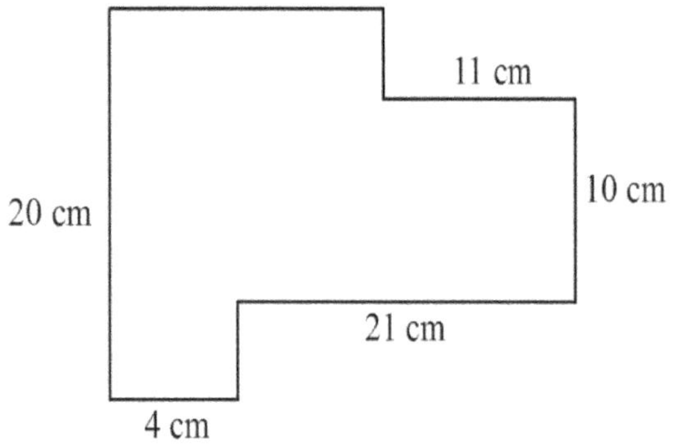

9.

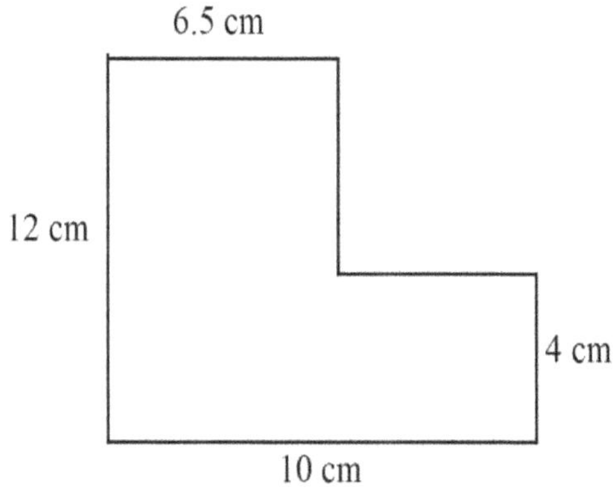

10.

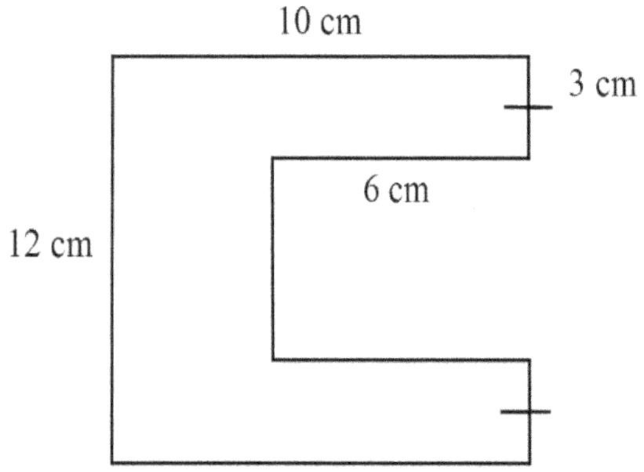

11.

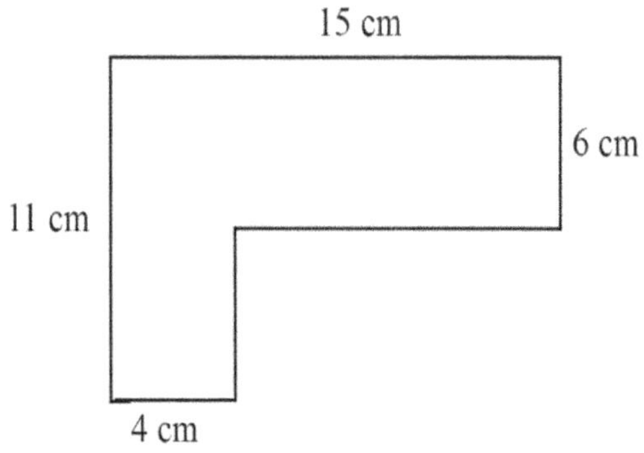

8B CIRCUMFERENCE OF A CIRCLE

The perimeter of a circle is called the **circumference**. We can calculate the circumference of a circle by using the formula $C = 2\pi r$, where r is the radius. We can also use $C = \pi d$, where d is the diameter.

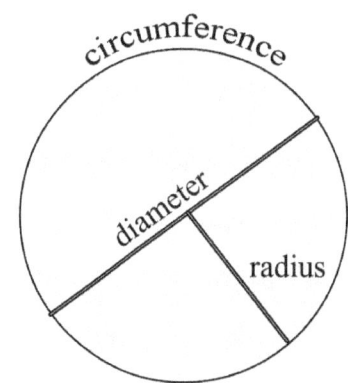

EXERCISE 8B

Find the circumference of each of the following circles.

1. 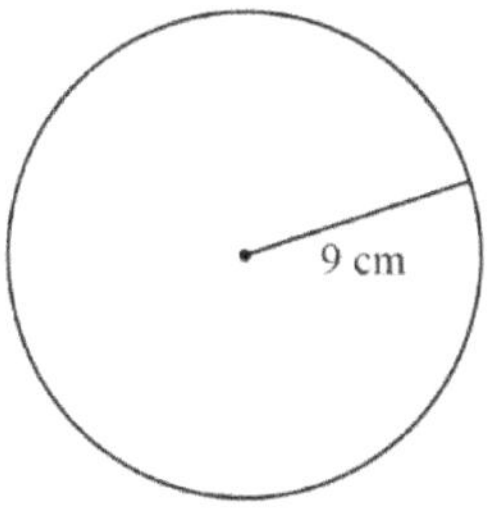 (circle with radius 9 cm) Using $C = 2\pi r$, $C = 2 \times \pi \times 9 = 56.55\ cm$	2. 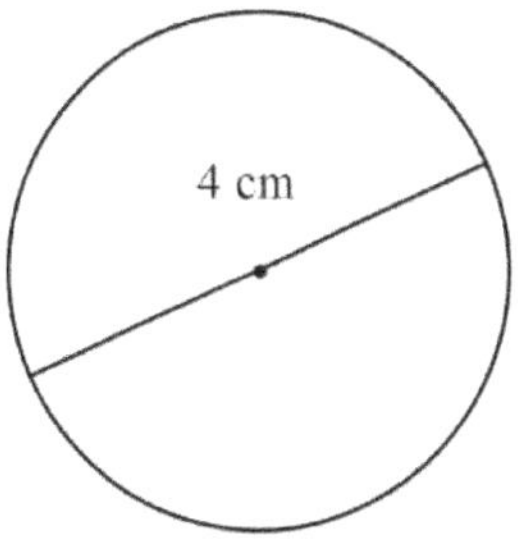 (circle with diameter 4 cm) Using $C = \pi d$, $C = \pi \times 4 = 12.57\ cm$
3. 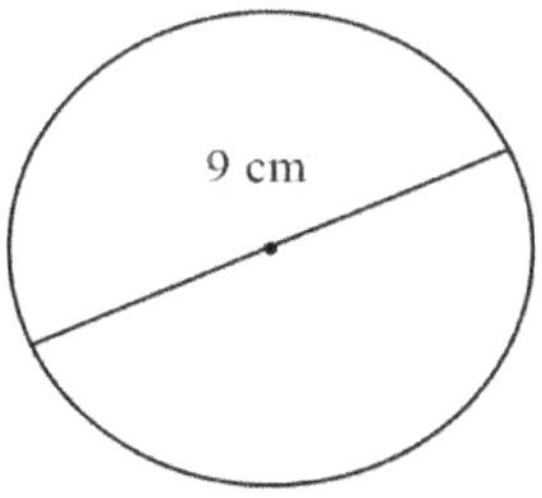 (circle with diameter 9 cm)	4. 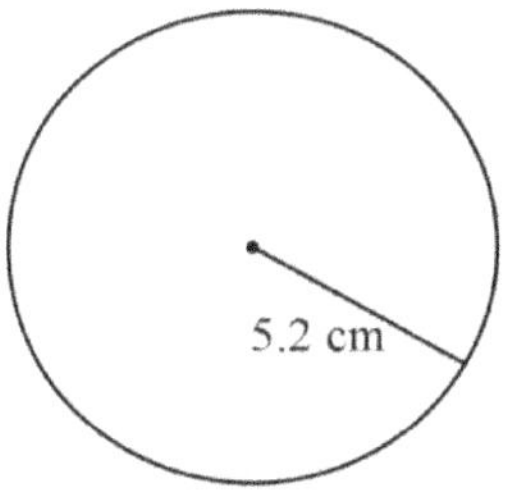 (circle with radius 5.2 cm)
5. (circle with radius 3.6 cm) 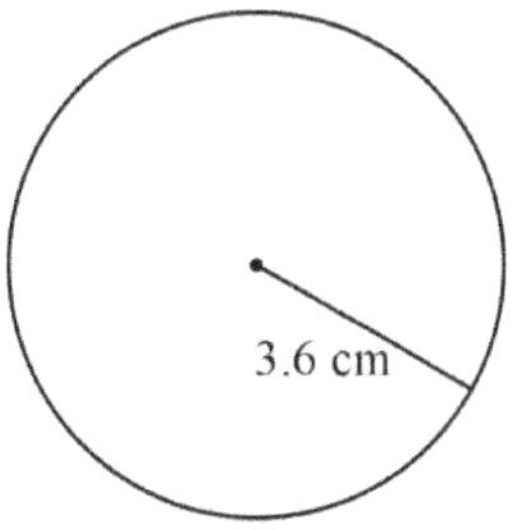	6. (circle with diameter 12.4 cm) 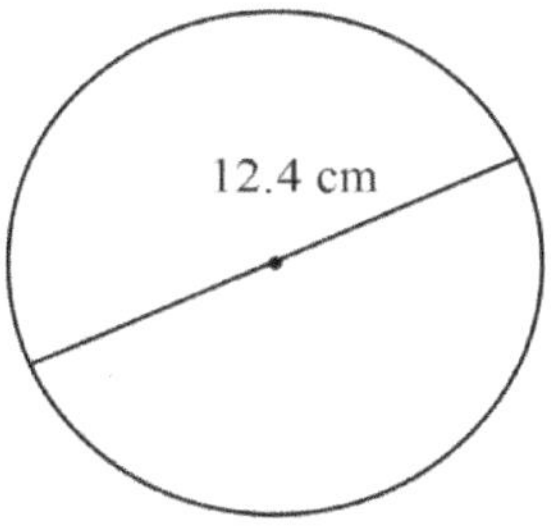
7. 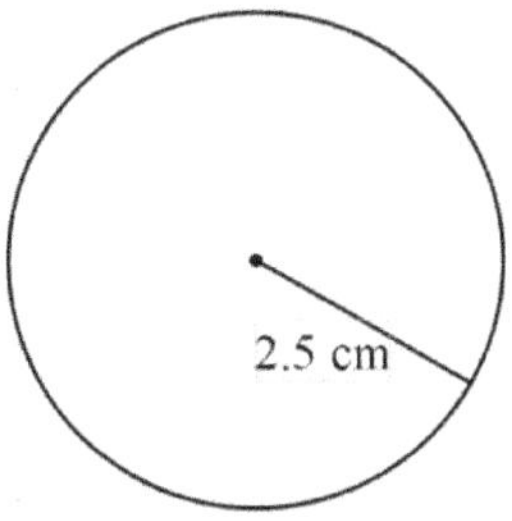 (circle with radius 2.5 cm)	8. 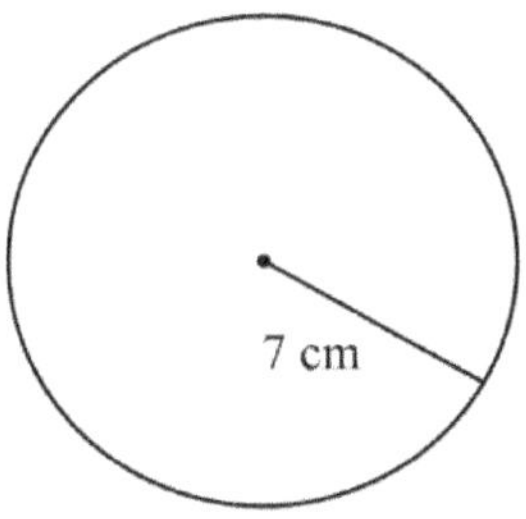 (circle with radius 7 cm)

8C ARC LENGTH AND PERIMETER OF A SECTOR

The boundaries of a sector of a circle A are formed by the two radii (OA and OB) and the arc (ADB).
A sector of a circle is just like a slice of pizza.

If we know the length of the radius and the angle subtended at the centre, we can find the length of arc and perimeter of the sector.

The arc length ADB can be found by using the formula

$$l = \frac{\theta}{360} \times 2\pi r,$$

where θ, measured in degrees, is the angle subtended by the arc ADB at the centre of the circle.

To find the perimeter of a sector of a circle, we need to add twice the length of the radius to the length of arc.

Perimeter of a sector is given by $\quad P = \frac{\theta}{360} \times 2\pi r + 2r \ or \ P = l + 2r.$

EXAMPLE

For the following sectors, find
    (i) arc length (ii) the perimeter.

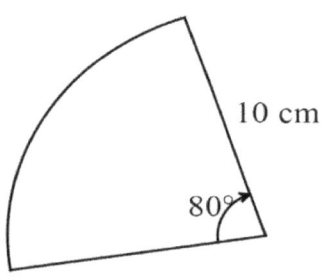

SOLUTION
(*i*) $arc\ length, l = \frac{\theta}{360} \times 2\pi r$

$= \frac{80}{360} \times 2 \times \pi \times 10$

$= 13.96\ cm$

(ii) Perimeter of minor sector OAB
$= l + 2r$
$= 13.96 + 2 \times 10$
$= 33.96\ cm$

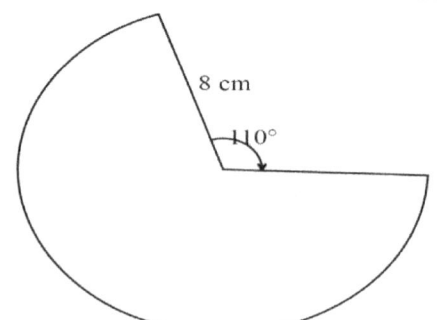

SOLUTION
(*i*) $arc\ length, l = \frac{\theta}{360} \times 2\pi r$
$\theta = 360 - 110 = 250$

$= \frac{250}{360} \times 2 \times \pi \times 8$

$= 34.91\ cm$

(ii) Perimeter of minor sector OAB
$= l + 2r$
$= 34.91 + 2 \times 8$
$= 50.91\ cm$

EXERCISE 8C

For the following sectors, find (i) arc length (ii) the perimeter.

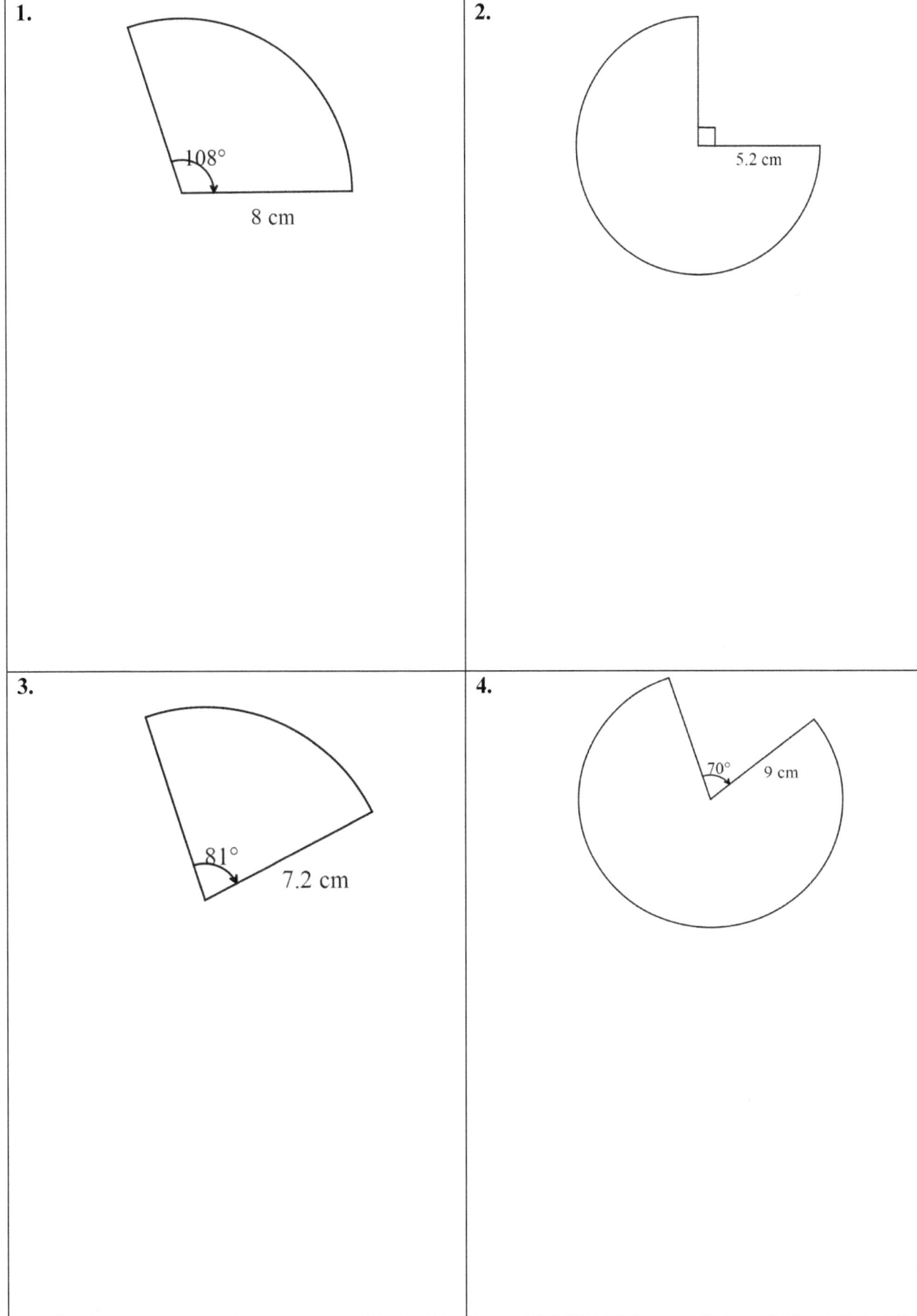

8D CONVERTING AREA UNITS

Area is simply the amount of surface within a particular boundary. Area is measured in square units such as cm², km² etc.. Because of the complexity of various units used in different situations, we have to be able to convert one unit to the other. The examples below will illustrate the whole process.

As area consists of square units, we have to square all our lengths.

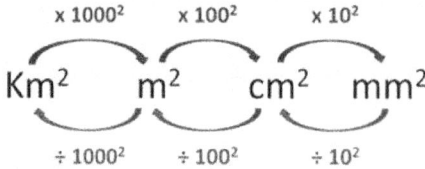

CLASS ACTIVITY

Complete the following conversions.

$6\ m^2 \to \cdots cm^2$ $6\ m^2 = 6 \times 100 \times 100$ $\qquad = 60000\ cm^2$	$1500\ mm^2 \to \cdots cm^2$ $1500\ mm^2 = 1500 \div 10 \div 10$ $\qquad = 15\ cm^2$
$5\ km^2 \to \cdots m^2$	$1200\ cm^2 \to \cdots m^2$
$13\ m^2 \to \cdots cm^2$	$12000\ mm^2 \to \cdots cm^2$
$500000\ m^2 \to \cdots km^2$	$7\ ha \to \cdots m^2\ (1\ ha = 10000\ m^2)$
$1400000\ mm^2 \to \cdots m^2$	$400\ ha \to \cdots km^2$

8E AREA FORMULAE

The table below shows different shapes and their respective area formula.

Square	Rectangle
$A = l^2$	$A = l \times w$
Parallelogram	**Trapezium**
$A = b \times h$	$A = \frac{a+b}{2} \times h$
Triangle	**Circle**
$A = \frac{1}{2}bh$	$A = \pi r^2$
Sector of a circle	**Semi-circle**
$A = \frac{\theta}{360} \times \pi r^2$	$A = \frac{\pi r^2}{2}$

EXERCISE 8E

Find the area of each shape.

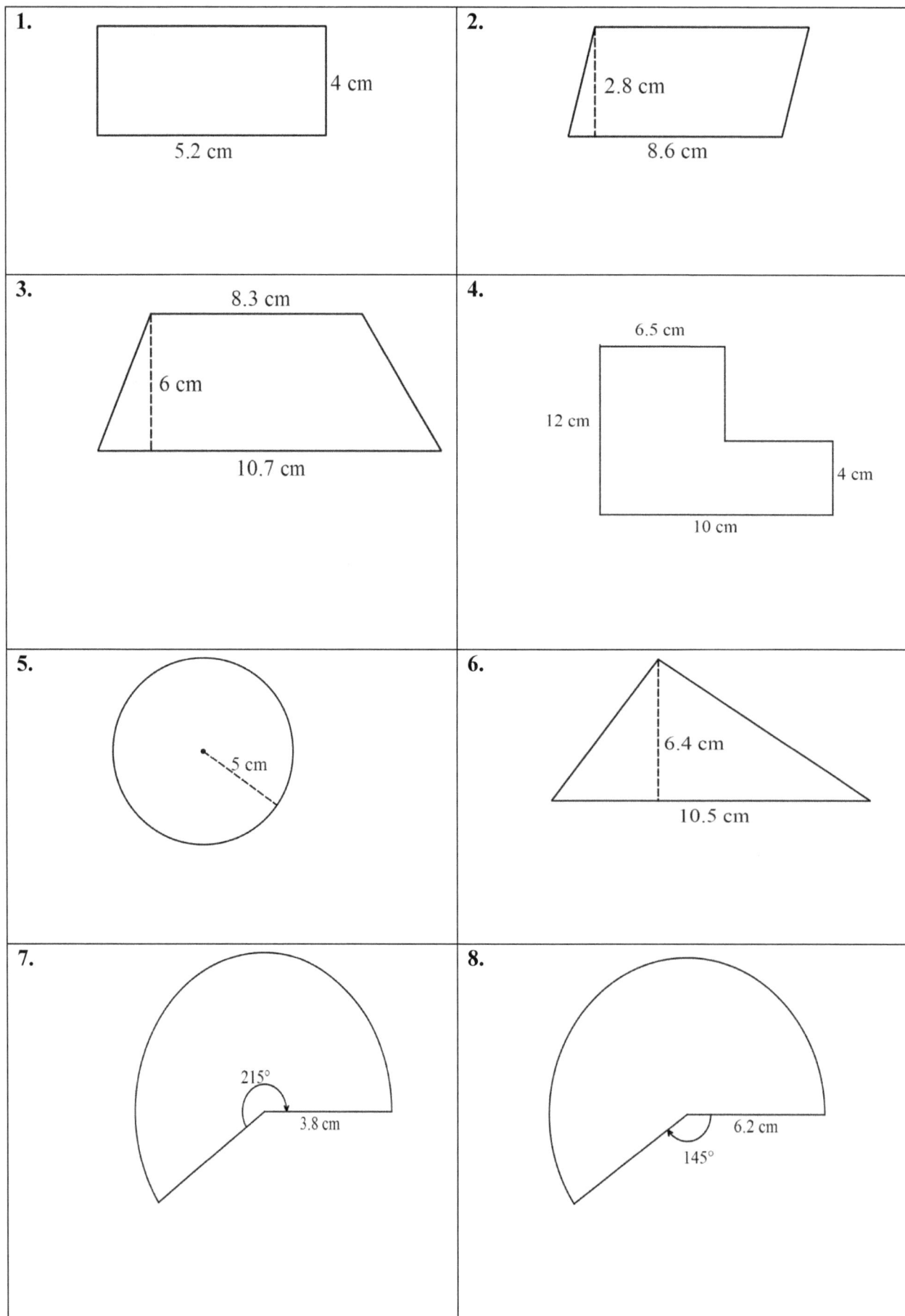

Find the shaded area of each of the following shapes. Give your answer to the nearest whole number.

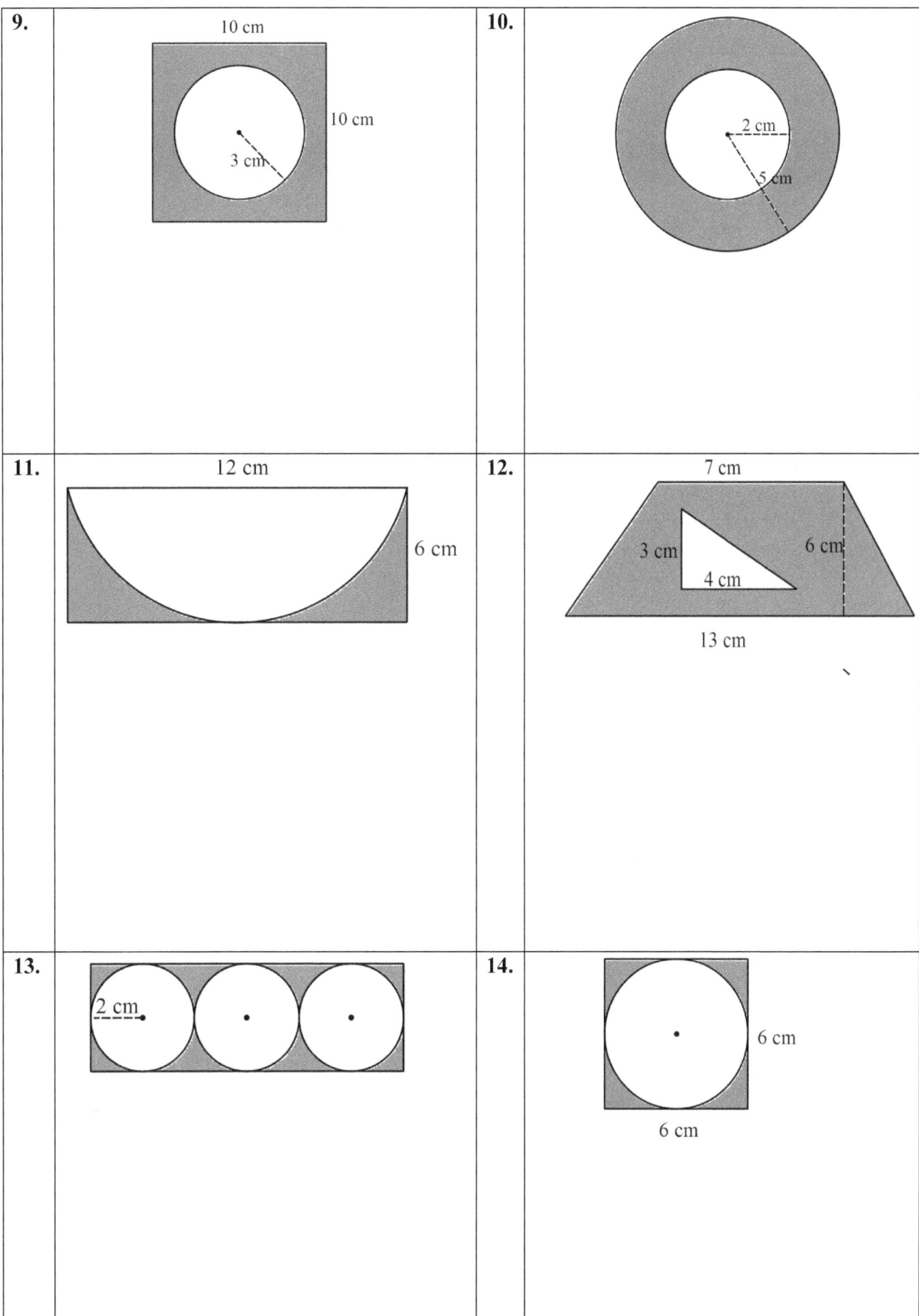

8F FINDING MISSING LENGTH AND RADIUS

In this part of the chapter we are going to find the length of missing sides in different shapes such as square, rectangle, triangle and so on given their respective areas. In addition, another objective would be to find the radius or circumference of a circle given its area.

EXAMPLES

1. A circle has an area of 25 cm². Find the radius of the circle. **SOLUTION** $$\pi r^2 = 25$$ $$r^2 = \frac{25}{\pi}$$ $$r = \sqrt{\frac{25}{\pi}} = 2.82 \; cm$$	2. A square has a perimeter of 40 cm. Find the area of the square. **SOLUTION** One side of the square = $40 \div 4 = 10 \; cm$ Area of square = $10 \times 10 = 100 \; cm^2$
3. A circle has a circumference of 24π cm. Find the area of the circle. **SOLUTION** $$2\pi r = 24\pi$$ $$r = 12$$ $$A = \pi(12)^2 = 452.39 \; cm^2$$	4. A triangle has a base length 12 cm and an area of 64.8 cm². Find the height of the triangle. **SOLUTION** $$\frac{1}{2}bh = 64.8$$ $$\frac{1}{2} \times 12 \times h = 64.8$$ $$6h = 64.8$$ $$\therefore h = 10.8 \; cm$$

EXERCISE 8F

1. A circle has an area of 36 cm². Find the radius of the circle.	2. A rectangle has a length of 11 cm. Given that the perimeter of the rectangle is 40 cm, find the width of the rectangle.
3. A circle has a circumference of $45.87 \; cm$. Find the area of the circle.	4. A circle has an area of 36π cm². Find the perimeter of the circle.

5. A square has a perimeter of 28 cm. Find the area of the square.	6. A trapezium has its pair of parallel sides as 12 cm and 8 cm respectively. Determine the height of the trapezium if it has an area of 96 cm².
7. A triangle has a base length 12.4 cm and an area of 62 cm². Find the height of the triangle.	8. A circle having diameter 6 cm has the same area as a square of side length x cm. Determine the value of x.

9. The following shapes have the same area. Determine the value of the pronumerals x, y and z.

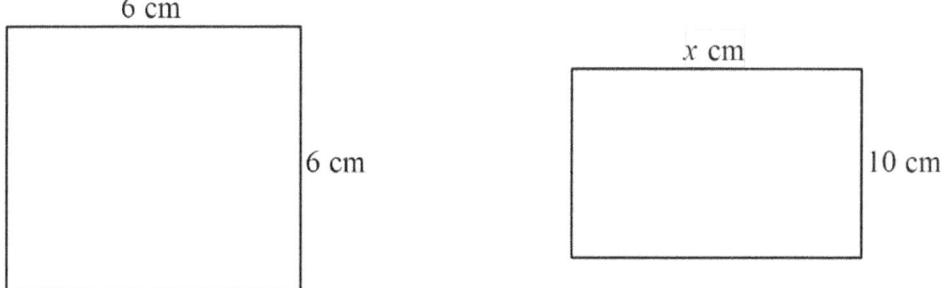

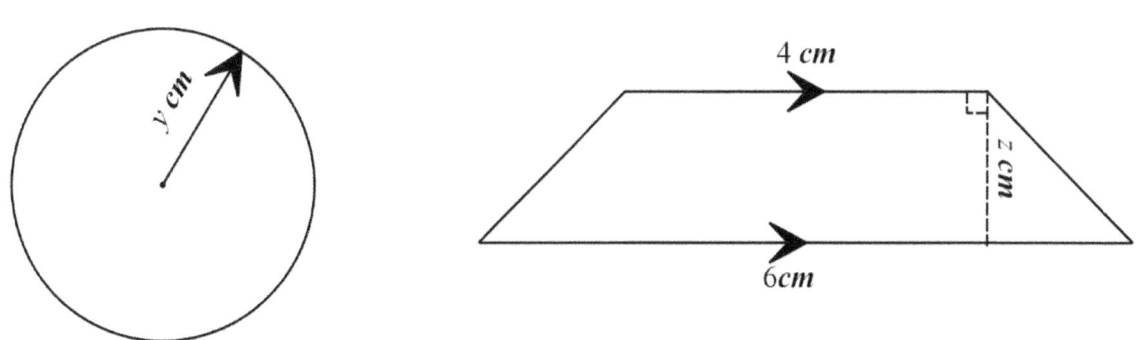

10. The following 3 shapes have the same area. Determine the missing sides marked with the letters a and x.

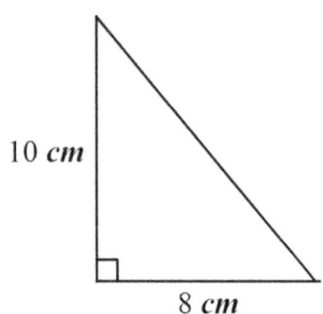

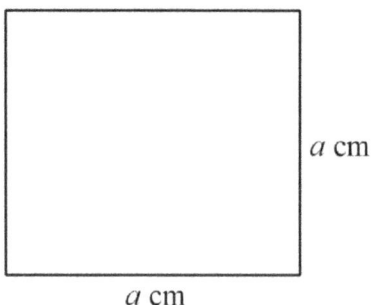

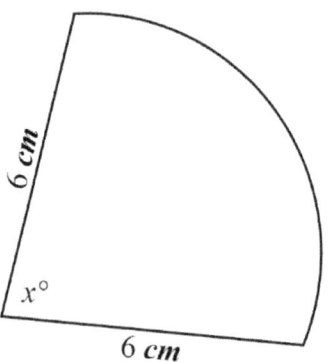

11. The following shapes have the same perimeter. Determine the values of x, y and z.

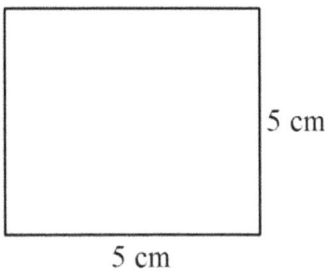

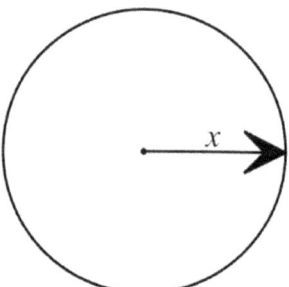

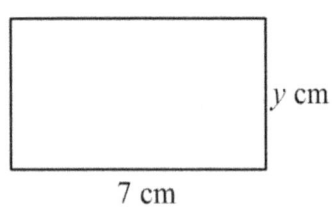

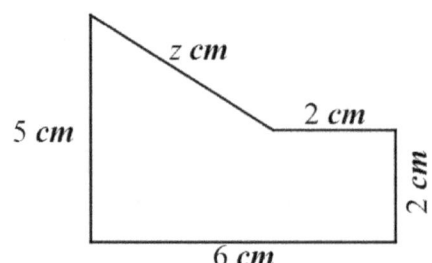

8G APPLICATIONS

1. The diagram shows part of Amy's earring.
   It is in the shape of a sector of a circle of radius 6 cm and angle 70°, from which a sector of radius 4 cm and angle 70° has been removed.
   Calculate the shaded area.

   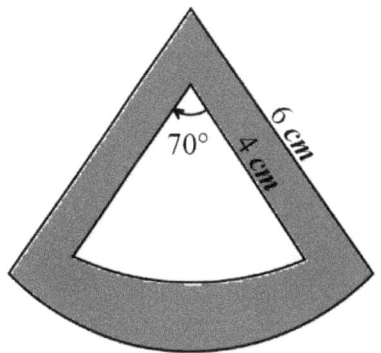

2. ABCD is a level field.
   F and E are points on AD such that BF and CE are perpendicular to AD.

   BF = 40 m and CE = 52 m.

   AF = 16 m, FE = 32 m and ED = 40 m.

   (a) Calculate the area of the field.

   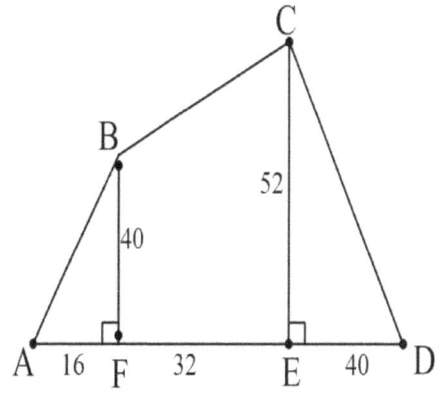

   (b) Calculate the length of BC.

3. AD and BC are arcs of circles with centre O.

A is a point on OB, and D is a point on OC.

OA = 10 cm and AB = 15 cm.

∠AOD = 130°.

(a) Calculate the perimeter of the shaded shape ABCD.

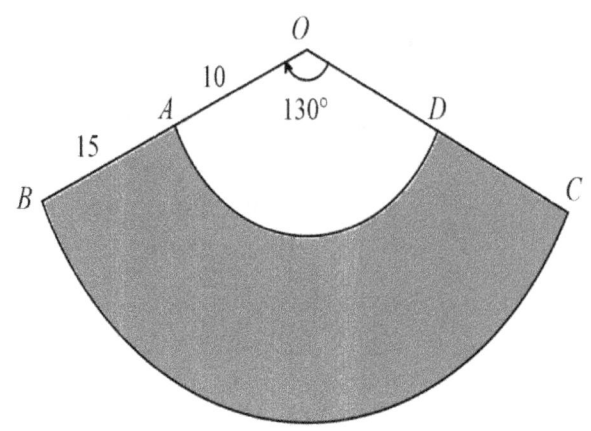

(b) Calculate the area of the shaded shape ABCD.

(c) The shape ABCD is used to make a lampshade by joining AB and DC.

Calculate the radius, r cm, of the circular top of the lampshade.

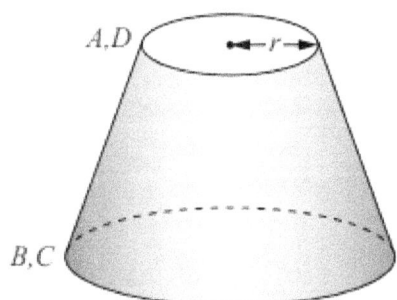

CHAPTER 8 : PERIMETER AND AREA

4. The Ryan's family recently bought a beach house in Rockingham. The house was built in the 1990's and as a result needs some renovation. As a priority, the family decided to tile all the four bedrooms as the carpet was old and their younger son being asthmatic was allergic to carpet. They had a few quotes and their best rate was $21.70 m² from ABC tilers.

(a) Calculate how much it would cost them in total to tile all the four bedrooms.

Bedroom 1: 4.2 m × 3.8 m

Bedroom 2 (trapezium): 4.9 m (top), 3.1 m (bottom), 2.9 m (height)

Bedroom 3: 5.2 m × 4.3 m with a 1.2 m × 1 m notch

Bedroom 4: circle with radius 2.5 m, centre O

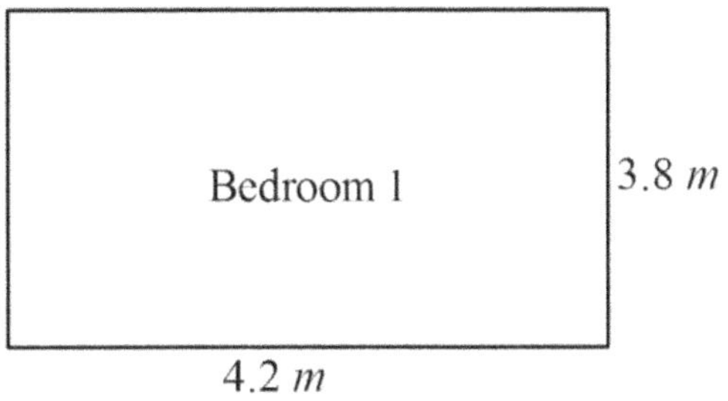

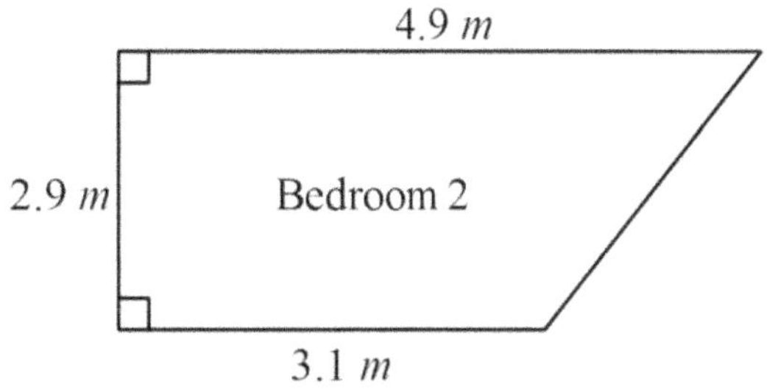

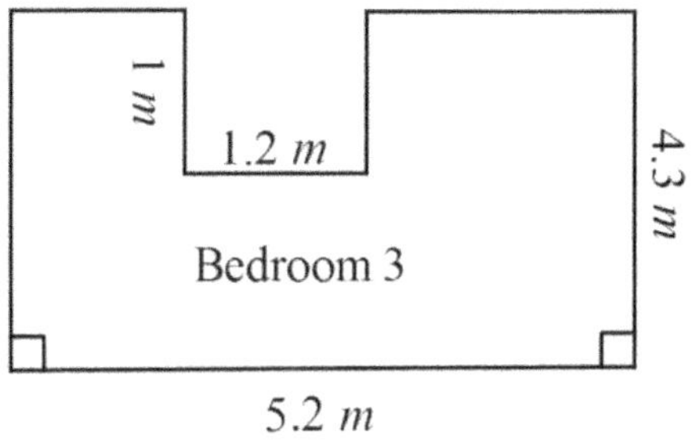

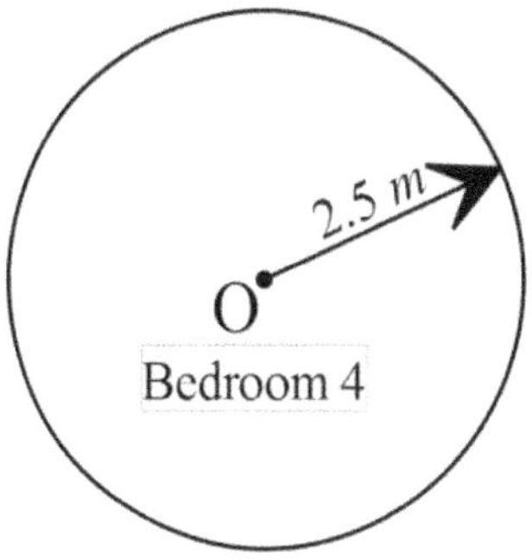

(b) Bedroom 4 being harder to tile, the Ryan's family was charged 5% extra for tiling that room. Determine the increase in cost.

5. Delta Pools are currently advertising 4 of its latest pools. The company is offering the Value collection pavers which are ideal for first home buyers or people looking to add a little beauty to their investment property.
Determine the cost to the nearest dollar of paving around each of the following pool areas.

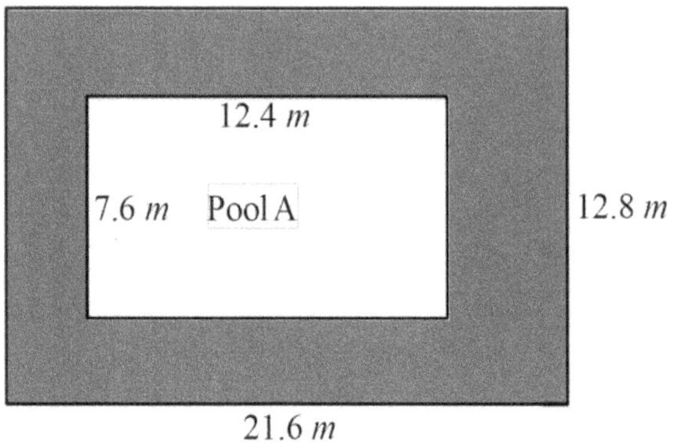

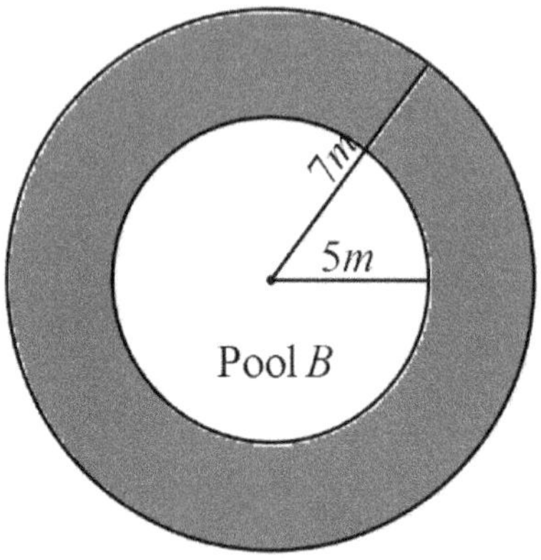

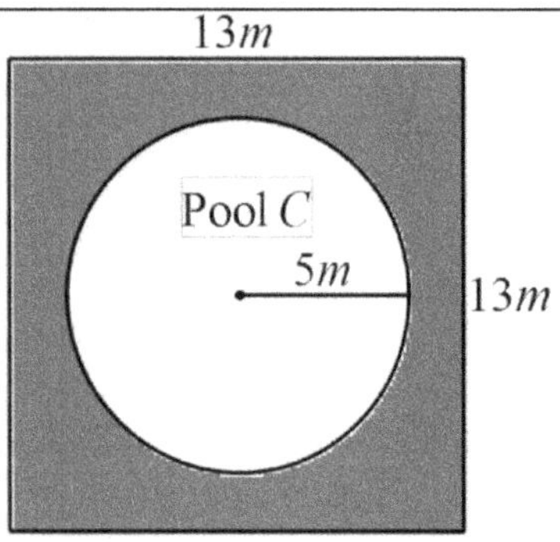

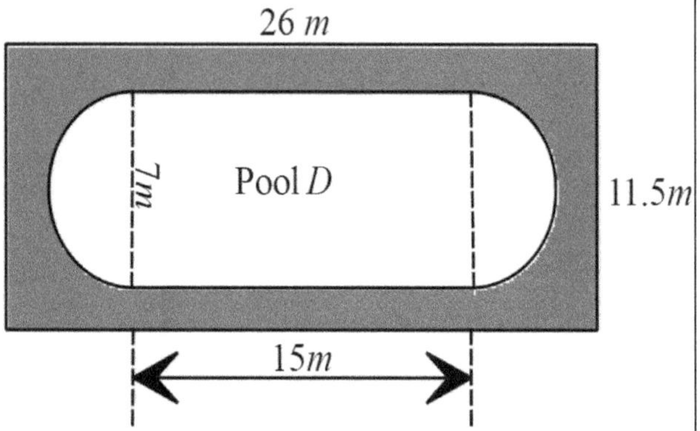

Pool A: 12.4 m × 7.6 m (inner); 21.6 m × 12.8 m (outer)

Pool B: inner radius 5 m, outer radius 7 m

Pool C: 13 m × 13 m square with circular pool of radius 5 m

Pool D: 26 m × 11.5 m rectangle with pool of width 7 m and straight section 15 m with semicircular ends

CHAPTER 9

SURFACE AREA AND VOLUME

9A SURFACE AREA AND VOLUME FORMULAE

Name of solid	Figure	Total surface Area	Volume
Cube		$A = 6x^2$	$V = x^3$
Rectangular prism		$A = 2(lw + wh + hl)$	$V = l \times w \times h$
Triangular Prism		$2 \times$ area of Δ + area of 3 rectangles	$V = \frac{1}{2}bh \times l$ or $V = $ Area of $\Delta \times l$
Cylinder		$A = 2\pi r(r + h)$ or $A = 2\pi r^2 + 2\pi rh$	$V = \pi r^2 h$
Sphere		$A = 4\pi r^2$	$V = \frac{4}{3}\pi r^3$
Cone		$A = \pi r(r + l)$ or $A = \pi r^2 + \pi rl$	$V = \frac{1}{3}\pi r^2 h$
Pyramid		Add base area + area of all Δs	$V = \frac{1}{3}$(Area of base) $\times h$

EXERCISE 9A

Find the total surface area and volume of each of the following solids. All units are in cm.

	Solids	Total surface area	Volume
1.	(cylinder, r=3, h=10)	$A = 2\pi r^2 + 2\pi rh$ $2 \times \pi \times 3^2 + 2 \times \pi \times 3 \times 10$ $= 245.04\ cm^2$	$V = \pi r^2 h$ $V = \pi \times 3^2 \times 10$ $= 282.74\ cm^3$
2.	(cylinder, diameter=21, height=8)		
3.	(cylinder, length=15, radius=3.5)		
4.	(cylinder, height=8, radius=2.5)		

CHAPTER 9 : SURFACE AREA AND VOLUME

	Solids	Total surface area	Volume
5.	(cuboid: 7.5 × 3 × 4)	$A = 2(lw + wh + hl)$ $2(7.5 \times 3 + 3 \times 4 + 4 \times 7.5)$ $= 129\ cm^2$	$V = l \times w \times h$ $V = 7.5 \times 3 \times 4 = 90\ cm^3$
6.	(cube: 8 × 8 × 8)	$A = 6x^2$ $A = 6 \times 8^2 = 384\ cm^2$	$V = x^3$ $V = 8^3 = 512\ cm^3$
7.	(cuboid: 12 × 5 × 6)		
8.	(cube: 9 × 9 × 9)		
9.	(cuboid: 10 × 4 × 7)		
10.	(cube: 6.1 × 6.1 × 6.1)		

	Solids	Total surface area	Volume
11.	(cone: height 8, radius 6)	Use Pythagoras to find the slant height. $l^2 = 6^2 + 8^2$ $\therefore l = 10$ $A = \pi r^2 + \pi r l$ $A = \pi \times 6^2 + \pi \times 6 \times 10$ $\quad = \quad cm^2$	$V = \frac{1}{3}\pi r^2 h$ $V = \frac{1}{3} \times \pi \times 6^2 \times 8$ $\quad = 301.59 \ cm^3$
12.	(cone: height 12, slant 13)		
13.	(cone: radius 5, height 24)		
14.	(cone: height 12, slant 15)		

	Solids	Total surface area	Volume
15.	(sphere, r = 4)	$A = 4\pi r^2$ $A = 4 \times \pi \times 4^2$ $= 201.06\ cm^2$	$V = \frac{4}{3}\pi r^3$ $V = \frac{4}{3} \times \pi \times 4^3$ $= 268.08\ cm^3$
16.	(hemisphere, r = 3.6)	$A = 3\pi r^2$ $A = 3 \times \pi \times 3.6^2$ $= 122.15\ cm^2$	$V = \frac{2}{3}\pi r^3$ $V = \frac{2}{3} \times \pi \times 3.6^3$ $= 97.72\ cm^3$
17.	(sphere, r = 4.5)		
18.	(hemisphere, r = 5)		
19.	(hemisphere, r = 3)		
20.	(sphere, r = 6.2)		

	Solids	Total surface area	Volume
21.	(triangular prism: triangle base 6, height 4, slant 5, length 18)	$2 \times$ area of Δ $= 2 \times \dfrac{1}{2} \times 6 \times 4 = 24$ Area of 3 rectangles $= 6 \times 18 + 18 \times 5 + 18 \times 5$ $= 288\ cm^2$ Total surface area $= 312\ cm^2$	$V =$ Area of $\Delta \times l$ $V = \dfrac{1}{2} \times 6 \times 4 \times 18$ $= 216\ cm^3$
22.	(triangular prism: base 16, height 6, slant 10, length 30)		
23.	(triangular prism: base 4, height 3, length 10)		
24.	(triangular prism: base 14, height 24, slant 25, length 50)		
25.	(triangular prism: slant 15, slant 20, height 9)		

	Solids	Total surface area	Volume
26.	(square pyramid: base 10, slant height 13, height 12)	Add base area + area of all △s Base area = 10 × 10 = 100 Area of 4 triangles = 4 × $\frac{1}{2}$ × 10 × 13 = 260 Total surface area = 360 cm²	$V = \frac{1}{3}$(Area of base) × h $V = \frac{1}{3}$ × 10 × 10 × 12 = 400 cm³
27.	(square pyramid: base 12, slant height 10, height 8)		
28.	(square pyramid: base 16, slant height 17, height 15)		
29.	(square pyramid: base ?, slant height 15, height 12)		
30.	(rectangular pyramid: base 6 cm × 6 cm, height 4 cm)		

9B FINDING RADIUS OR SIDE

In this section of the chapter, the volume of different shapes would be given and our aim would be to find the radius or missing sides from cylinders, cones, spheres, cube and so on.

EXAMPLES

In the examples below, we can alternatively make use of the solve facility of the CAS calculator to find the missing values.

1. A sphere has a volume of 400 cm³. Find the radius of the sphere correct to one decimal place. **SOLUTION** $$\frac{4}{3}\pi r^3 = 400$$ $$r^3 = \frac{400 \times 3}{4\pi}$$ $$r = \sqrt[3]{\frac{1200}{4\pi}} = 4.6\ cm$$	2. A cube has surface area of 150 cm². Find the volume of the cube. **SOLUTION** $$6x^2 = 150$$ $$x^2 = 25$$ $$\therefore\ x = 5$$ Volume of cube = $5^3 = 125\ cm^3$
3. A cylinder has a radius of 6 cm and a volume of 1244.07 cm³. Determine the height of the cylinder to the nearest cm. **SOLUTION** $$\pi r^2 h = 1244.07$$ $$\pi \times 6^2 \times h = 1244.07$$ $$h = \frac{1244.07}{36\pi} = 11\ cm$$	4. A right circular cone has a volume of 442.44 cm³. Given that its perpendicular height is 10 cm, determine the radius of the cone. **SOLUTION** $$\frac{1}{3}\pi r^2 h = 442.44$$ $$\frac{1}{3}\pi \times r^2 \times 10 = 442.44$$ $$r^2 = \frac{442.44 \times 3}{10\pi}$$ $$r = \sqrt{\frac{442.44 \times 3}{10\pi}} = 6.5\ cm$$

5. A cube of side 5 cm and a sphere of radius 6 cm are both melted down to form a cylinder of height 10 cm. Determine the radius of the cylinder.
   SOLUTION
   Volume of cube = $5^3 = 125\ cm^3$
   Volume of sphere = $\frac{4}{3} \times \pi \times 6^3 = 904.78\ cm^3$
   Total volume = 1029.78
   CAS (Solve ($\pi \times r^2 \times 10 = 1029.78, r$))
   $r = 5.73\ cm$

EXERCISE 9B

1. A sphere has a volume of 360 cm^3. Find the radius of the sphere correct to one decimal place.

2. A cube has a volume of 216 cm^3. Find the total surface area of the cube.

3. A cylindrical tank of height 40 cm and radius r cm has a capacity of 18 litres. Find the radius to the nearest centimetre. (1 litre = 1000 cm^3)

4. The cylinder on the right has a radius of 4 cm and height h cm. The volume of the cylinder is 1200 cm^3. Find h.

5. A right circular cone has a volume of 603.19 cm^3. Given that its radius is 8 cm, determine the perpendicular height of the cone.

6. A cube of side 10 cm is melted down to form a cylinder of height 8 cm. Determine the radius of the cylinder.

7. The following six shapes have the same volume. Determine each of the pronumerals marked x, y, z, w and m.

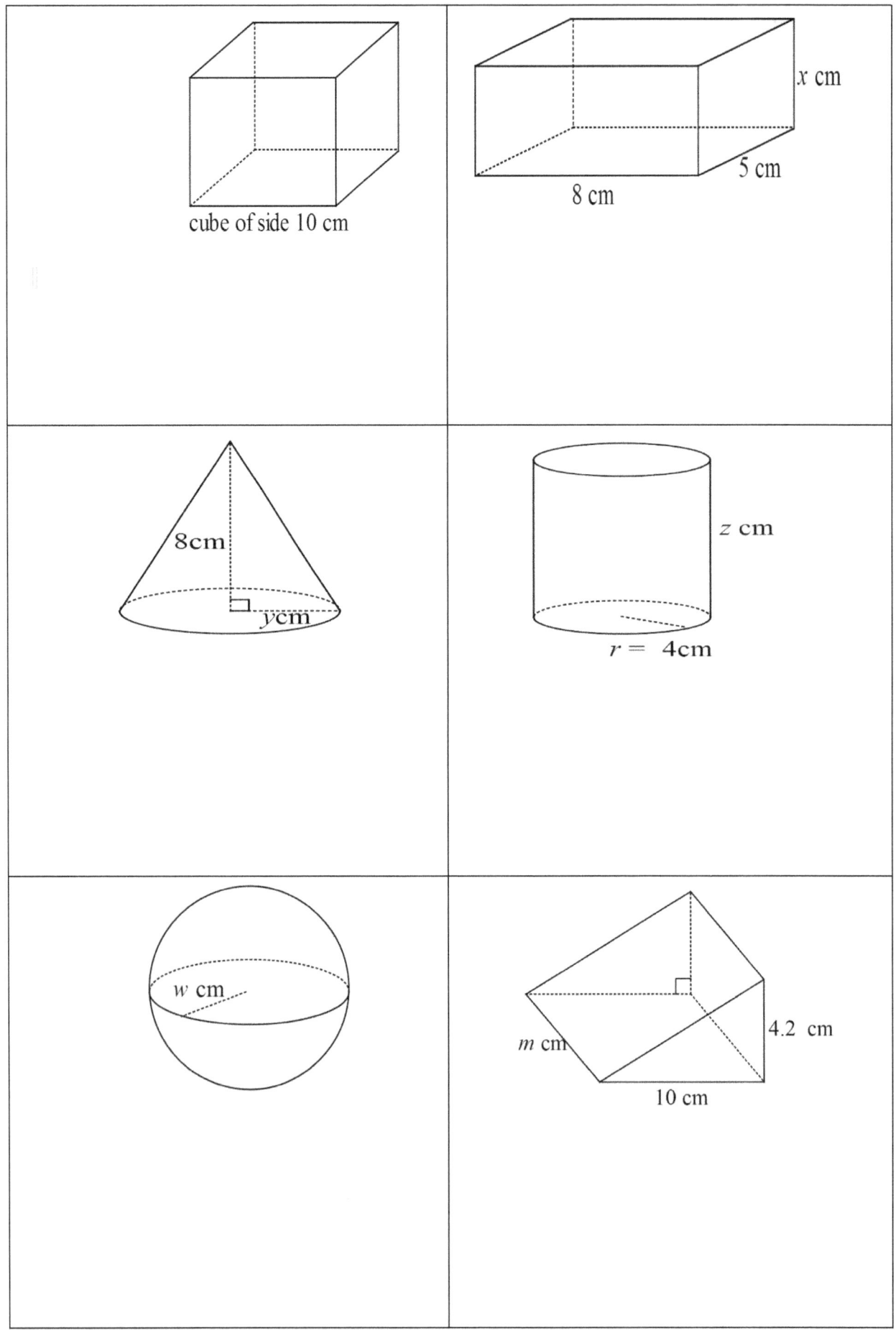

8. The following four shapes have the same total surface area. Determine each of the pronumerals marked *a, h* and *x*.

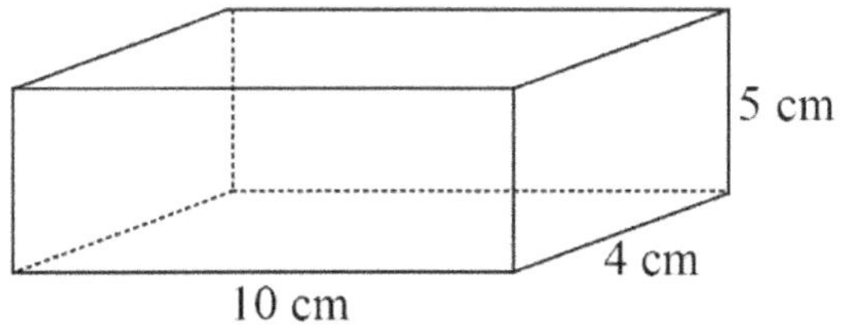

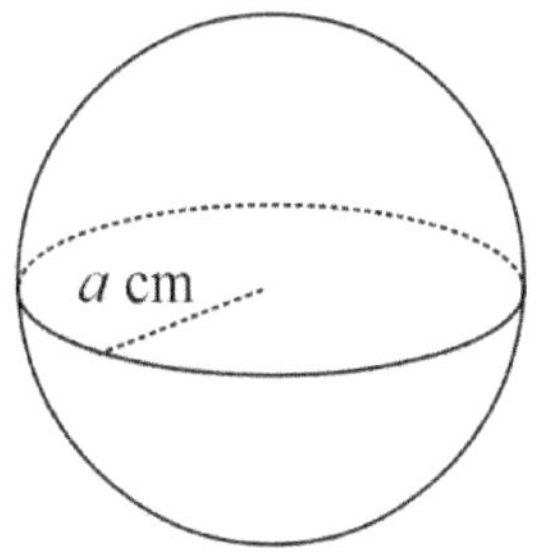

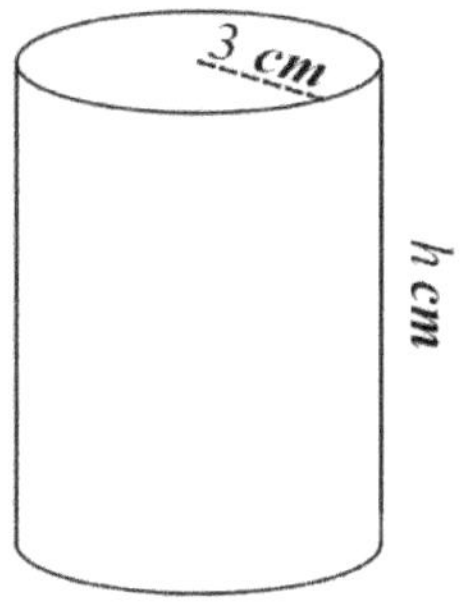

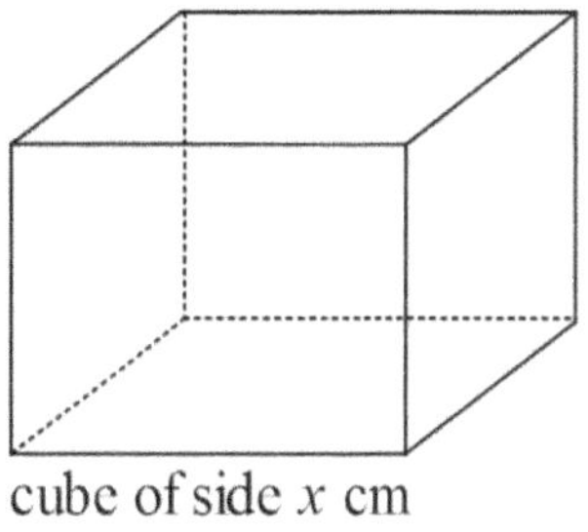

5 cm, 4 cm, 10 cm (rectangular prism)	*a* cm (sphere)
3 cm, *h* cm (cylinder)	cube of side *x* cm

9. The solid below have a total surface area of 162 cm³. Determine the value of *x*.

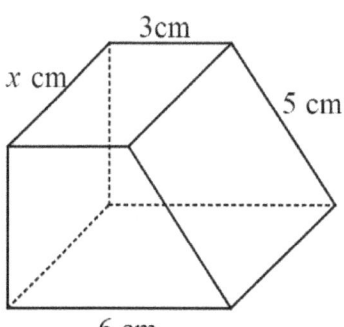

9C CAPACITY

Capacity is the volume of liquid that can fit inside a container. For example, the capacity of a carton of milk is the amount of milk that can fit inside the carton.

The standard units for capacity are the litre (L) and millilitre (mL). Note the following:

$$1 \, mL = 1 \, cm^3$$
$$1 \, litre = 1000 \, mL = 1000 cm^3$$
$$1 \, kL = 1000 L$$
$$1 ML = 1\,000\,000 \, L$$

CLASS ACTIVITY

Complete the following table

1. $50 \, cm^3 = $ _____ mL	2. $4500 \, mL = $ _____ L
3. $4 \, L = $ _____ mL	4. $3.4 \, L = $ _____ cm^3
5. $7000 \, cm^3 = $ _____ L	6. $3.65 \, kL = $ _____ L
7. $33.7 \, L = $ _____ cm^3	8. $3850 \, L = $ _____ kL
9. $900 \, mL = $ _____ cm^3	10. $2.54 \, L = $ _____ cm^3

11. A fish tank is in the shape of a rectangular prism. It is 65cm long, 40 cm wide and 50cm high. Determine the capacity of the tank.

12. Find the capacity of the container on the right.

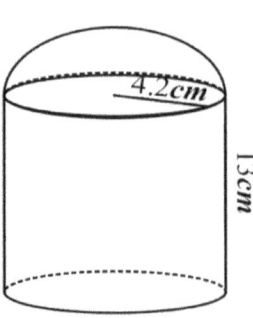

13. The following three cylindrical containers are filled with a certain liquid. Rank them according to their capacity, giving your answer in descending order.

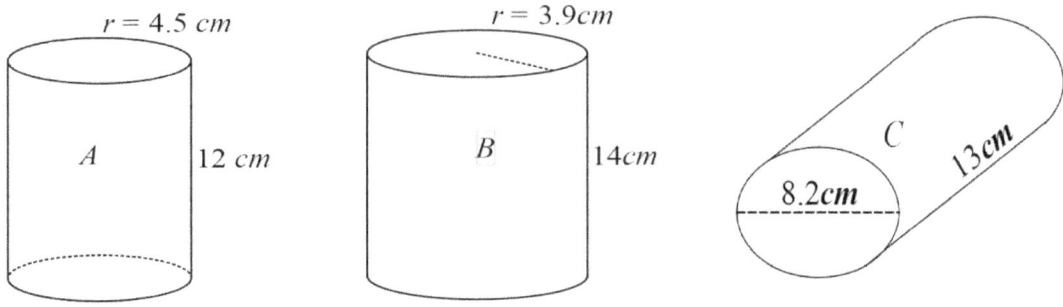

14. Three cartons P, Q or R are filled with milk. Which of them has the greatest capacity? Show your workings.

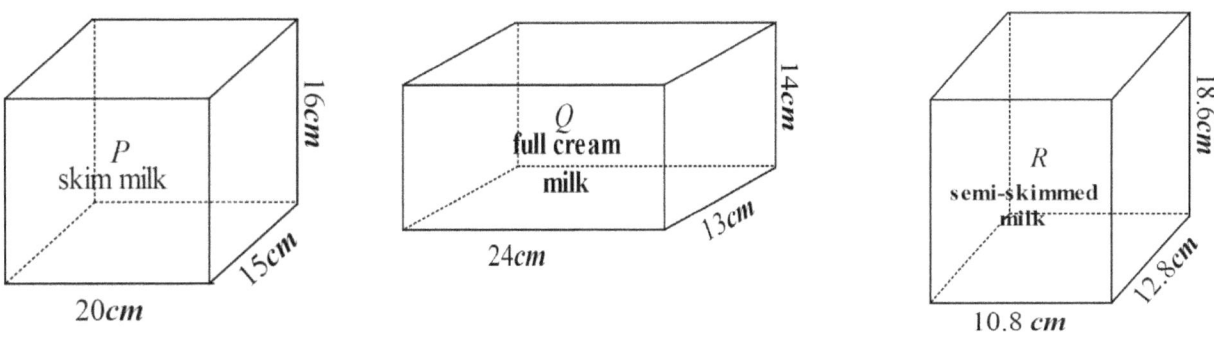

9D VOLUME OF OTHER PRISMS

The volume of a prism is given by the formula:

Volume of prism = Area of cross section × length

EXAMPLES

Find the volume of the prism.	The volume of the prism is 182.70 m². Find the area of the cross section.
Area of cross section = 20m², length = 40m $Volume = 20 \times 40 = 800 m^3$	length = 14.5m $Area = 182.70 \div 14.5 = 12.6 m^2$

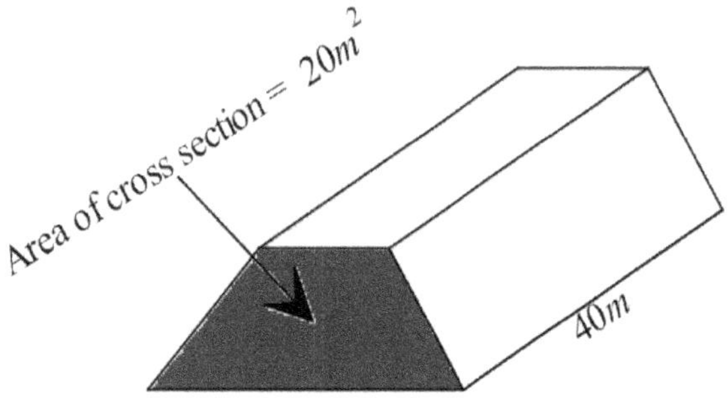

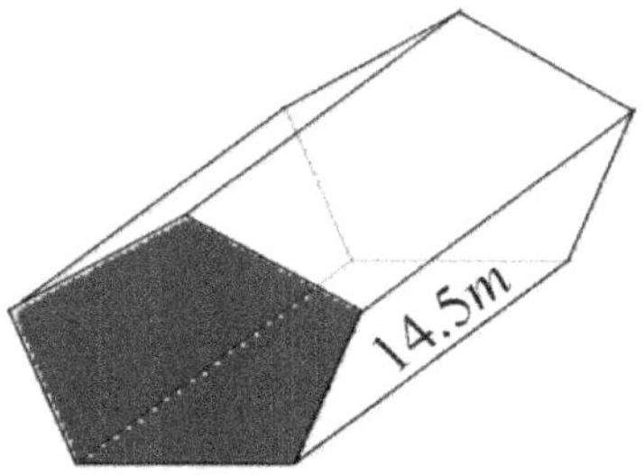

EXERCISE 9D

1. Find the volume of the prism.

   26m, $A = 32m^2$

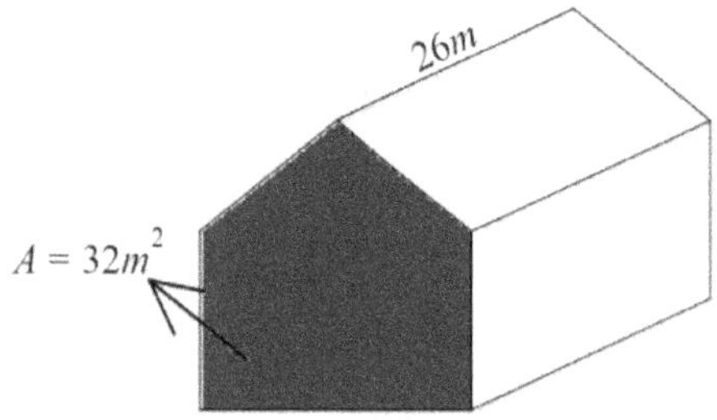

2. Find the volume of the prism.

   12.8m, Area of end = 10.2m²

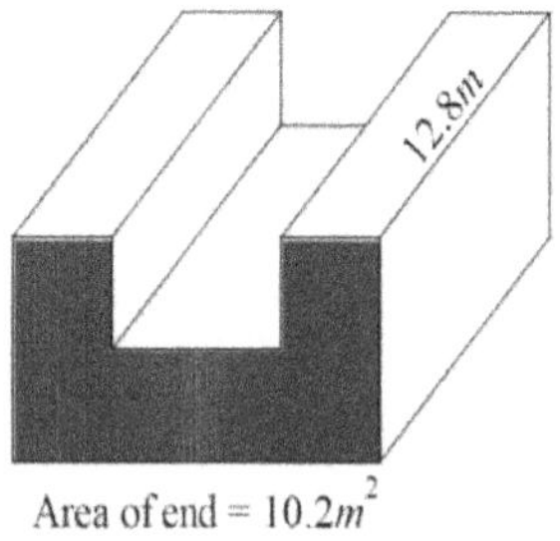

3. Find the volume of the prism.

   21.5m, $A = 28m^2$

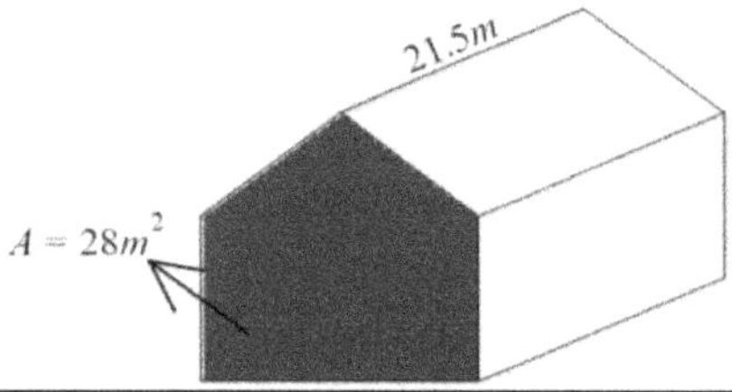

4. The volume of the prism is 259.59 m². Find the area of the cross section.

   25.45m

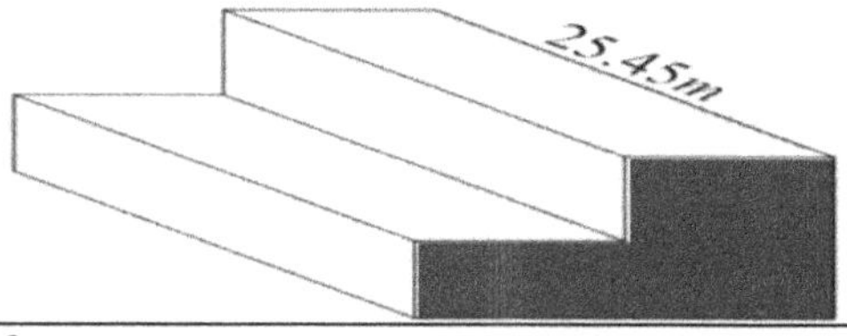

5. The volume of the prism is 365.20 cm³. Find the value of x.

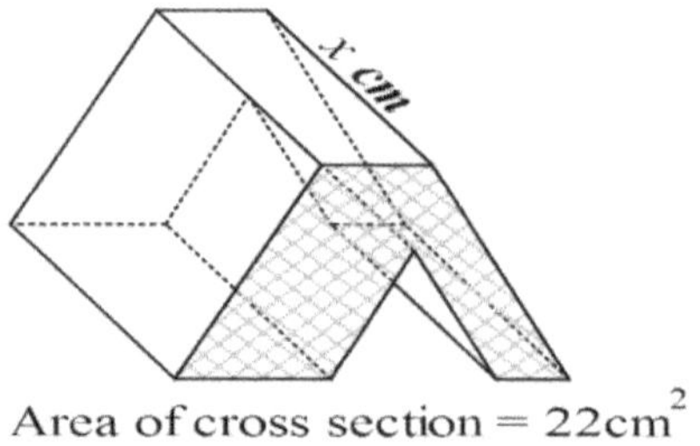

   x cm, Area of cross section = 22cm²

9E APPLICATIONS

1. Sheldon made a bookcase during his woodwork class to keep his comic books.
His teacher gave him a 40 cm long, 20 cm wide and 8 cm high rectangular wooden block.
He carved out an 18 cm by 4 cm rectangular hole in the block.
Calculate the volume of the bookcase.

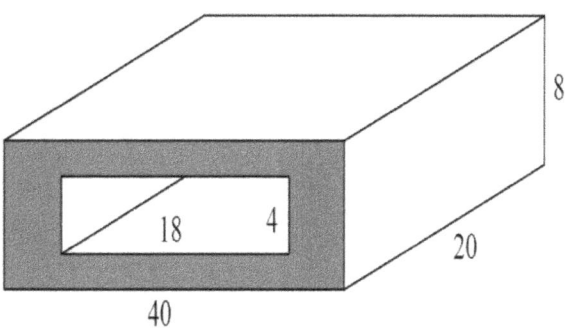

2. The diagram represents a solid block of wood of length 20m. The faces ABCD and EFGH are horizontal rectangles. The faces ABFE, BCGF and ADHE are vertical.
BC = AD = 12m, BF = AE = 8m and FG = EH =18m.

Calculate

(a) the length of CG,

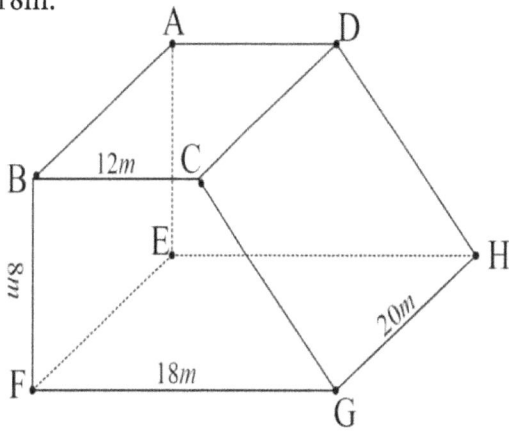

(b) the volume of the block,

(c) the total surface area of the block.

3. The diagram shows a wine glass. The base is a hemisphere of radius 3 cm and the top part is a cone of radius 4cm and vertical height 15 cm.

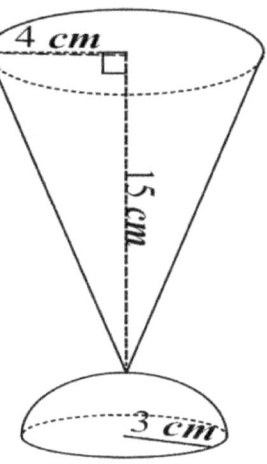

(a) Determine the volume of the whole container.

(b) Peter poured 280 mL of wine in the glass. Would the wine overflow?

4. Teddy toad lives in a castle made from mud and sticks under the water. He wants to start a family so he needs to work out how many little taddies he can have. Each one will require a space of 5 cm^3. He needs at least 150 cm^3 for himself. Calculate how many taddies he can have.

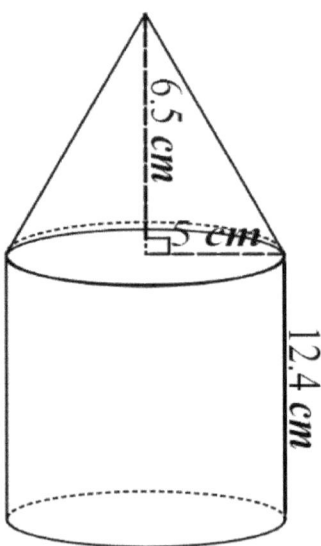

5. A water tanker delivers water in a cylindrical container of length 18 m and radius 1.5 m. After several deliveries, the water remaining in the container is shown in the diagram.

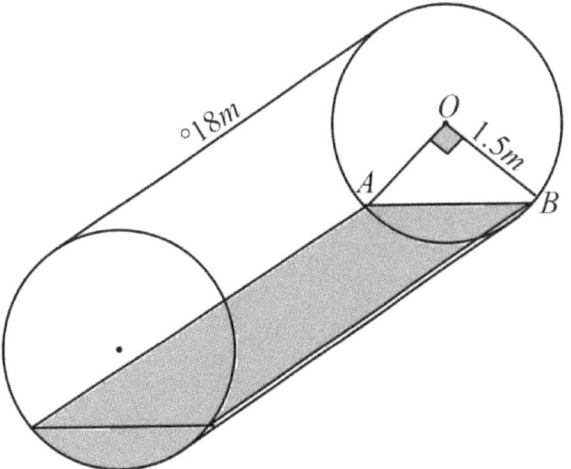

AB is horizontal, O is the centre of the circular cross-section and ∠AOB = 90°.

(i) Calculate the area of the minor sector OAB, to three decimal places.

(ii) Calculate the area of ΔOAB,

(iii) Hence calculate the curved surface area of the container that is in contact with the water.

(iv) Calculate the volume of water remaining in the container.

6. The diagram on the right shows a garage shed as advertised in the yellow pages. As an investigation, Mr Thompson, the Mathematics teacher, has assigned his students to work out

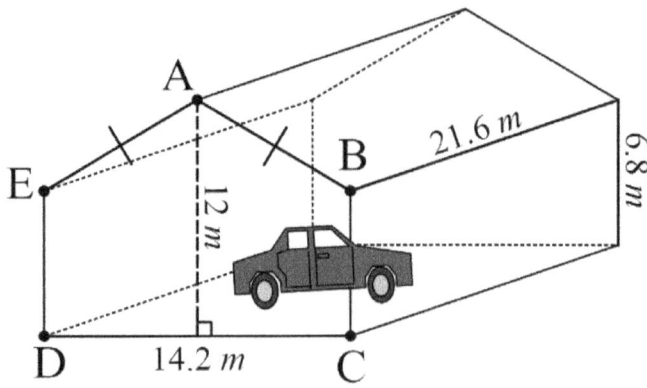

(a) the length of AB,

(b) Area of the face ABCDE,

(c) the volume of the shed.

(d) the total surface area of the shed.

CHAPTER 10

SIMILARITY

10A SIMILAR FIGURES AND SIMILAR TRIANGLES

Similar figures have identical shape, but differ in size. Similar figures are different from congruent figures. The latter have the same shape and are exactly the same size.

The calculators shown below are similar. Clearly, the length and width of the smaller calculator has been multiplied by the same scale factor of 2 to produce the length and the width of the larger calculator.

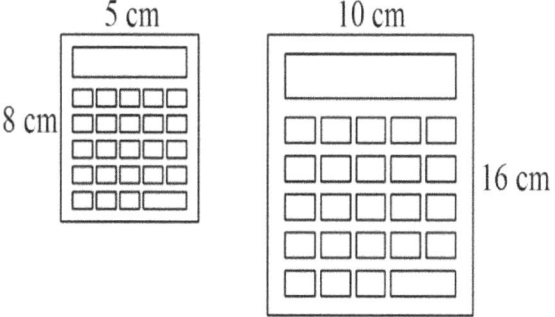

Consider the triangles below. They are obviously similar.

Use the angles to match each corresponding sides.

$\dfrac{EF}{CB} = \dfrac{6}{4} = 1.5$

$\dfrac{DF}{AC} = \dfrac{9}{6} = 1.5$

$\dfrac{DE}{AB} = \dfrac{4.5}{3} = 1.5$

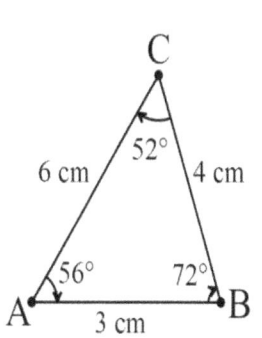
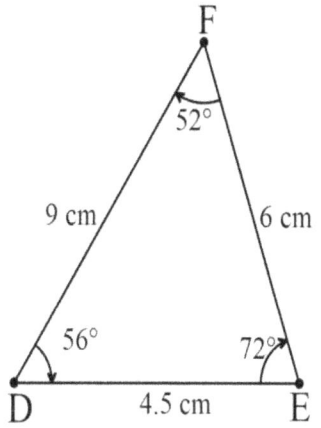

In the triangles ABC and DEF, matching angles are equal as well as the sides have equal ratios (scale factor). Hence, we can say that $\triangle$ ABC is similar to $\triangle$ DEF.

Symbolically, we write $\triangle$ ABC ~ $\triangle$ DEF.

As a conclusion, in similar figures

- All corresponding sides are in the same ratio
- All pairs of corresponding angles are equal.

10B TESTING SIMILAR TRIANGLES

To prove two triangles are similar, we do not need to know all the sides and all the angles. The four tests listed below can be used to show that two triangles are similar. Depending on what information is given, we have to choose the appropriate test.

1 Angle, Angle, Angle (AAA)

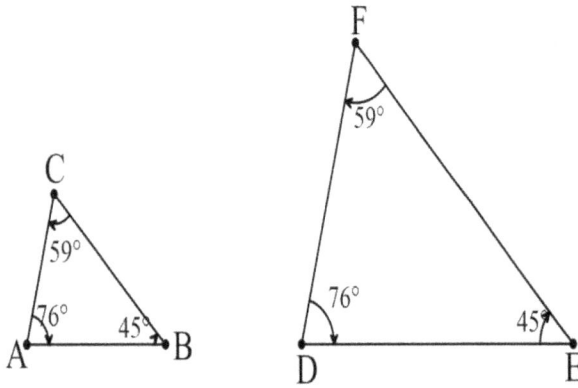

Since three pairs of angles are of the same magnitude,
△ ABC ~ △ DEF (AAA).
It any two angles are given in a triangle we can work out the third angle as angle sum of a triangle is 180°.

2 Side, Side, Side (SSS)

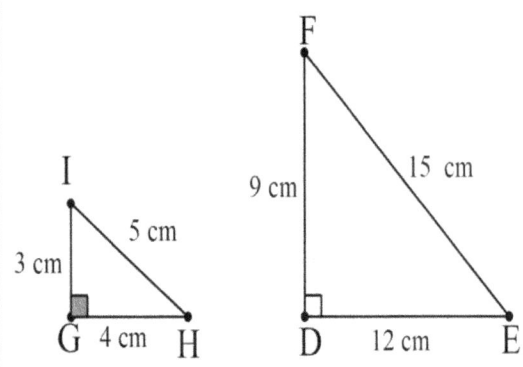

Since three pairs of matching sides are in the same ratio, scale factor being 3.
△ GHI ~ △ DEF (SSS).

3 Side, Angle, Side (SAS)

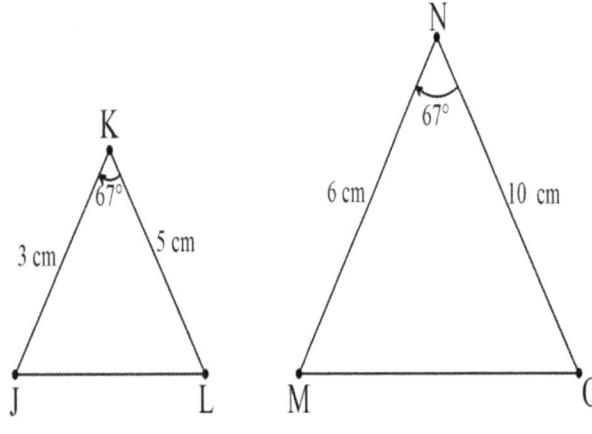

Since two pairs of corresponding sides are in the same ratio and the angle included between the two sides are equal, we say
△ JKL ~ △ MNO (SAS)

4 Right Angle, Hypotenuse, Side (RHS)

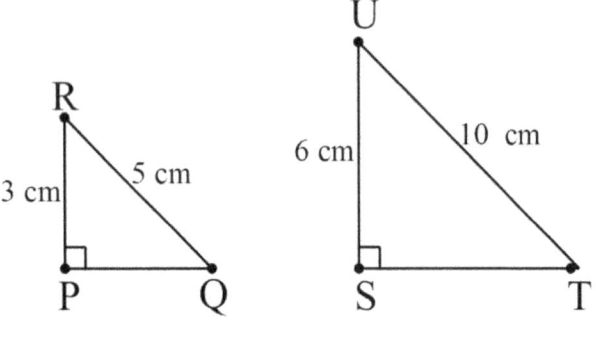

Since the hypotenuse and one side in △PQR is in the same ratio as the hypotenuse and one side of △STU, we can say that
△ PQR ~ △ STU (RHS)

EXAMPLE

Use similarity test to show that the following triangles are similar.

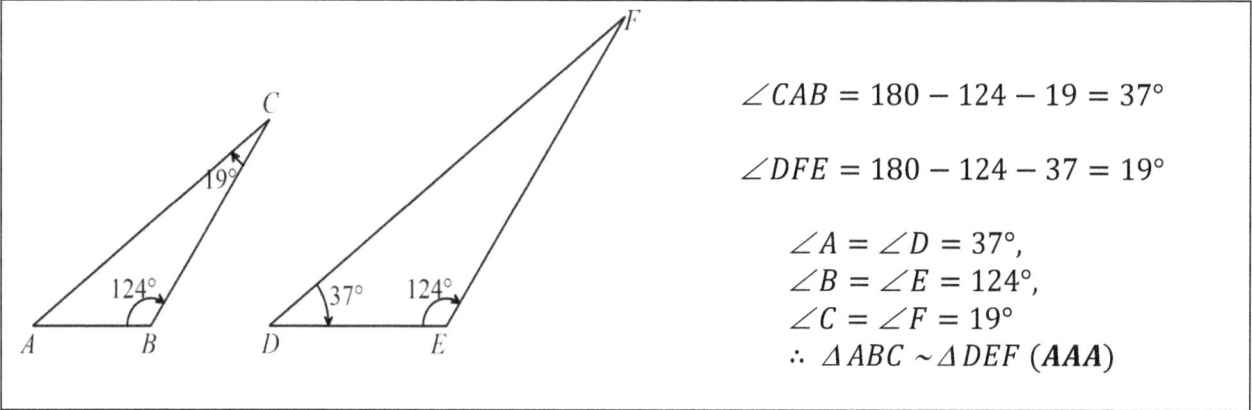

$\angle CAB = 180 - 124 - 19 = 37°$

$\angle DFE = 180 - 124 - 37 = 19°$

$\angle A = \angle D = 37°,$
$\angle B = \angle E = 124°,$
$\angle C = \angle F = 19°$
$\therefore \triangle ABC \sim \triangle DEF$ (**AAA**)

EXERCISE 10B

Use similarity test to show that the following triangles are similar.

1.

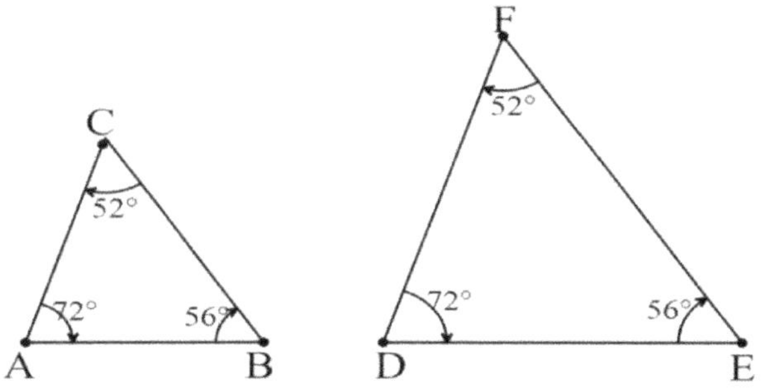

2.

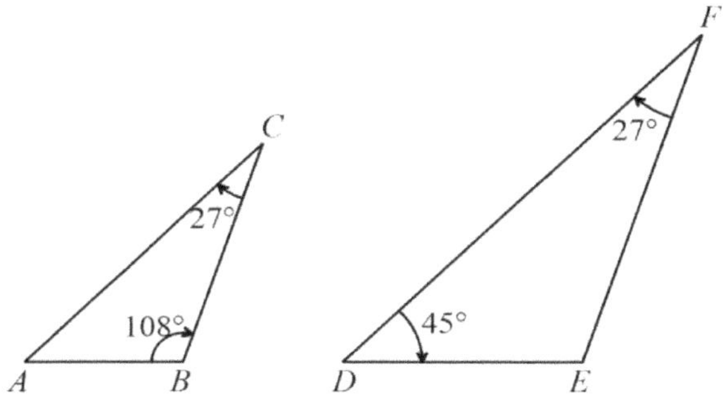

3.

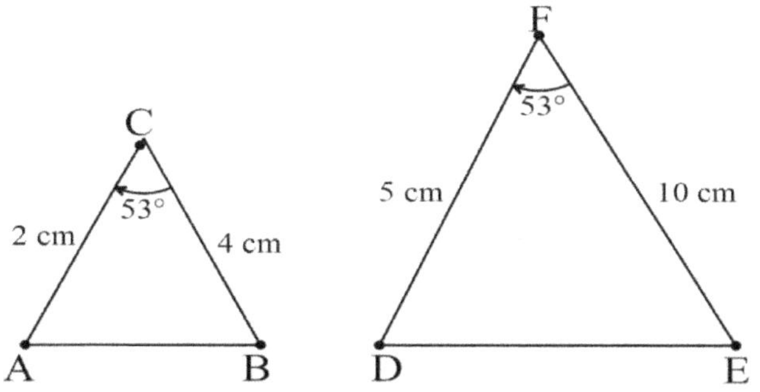

Use similarity test to show that the following triangles are similar.

4.

5.

6.

7.

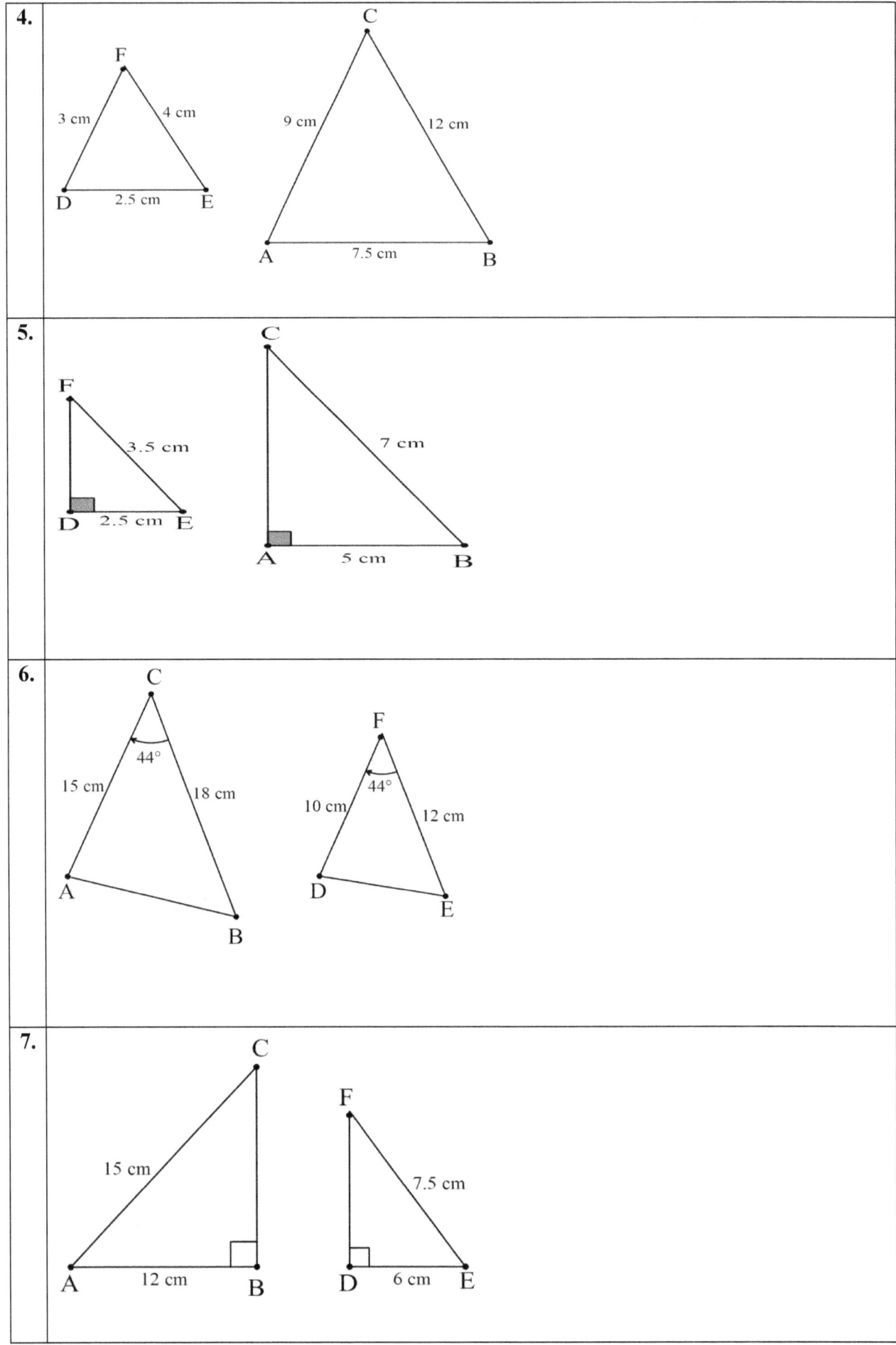

CHAPTER 10 : SIMILARITY

10C SIMILARITY AND SCALE FACTOR

If two triangles are similar, the scale factor is the ratio of the corresponding sides.
In the triangles on the right,

- The side corresponding to RQ is UT
- The side corresponding to PR is SU

$$\frac{PR}{SU} = \frac{RQ}{UT} = \frac{1}{2}$$

∴ $\triangle PQR$ is half the size of $\triangle STU$ in terms of length.

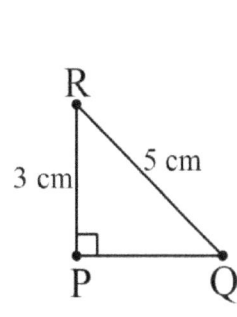
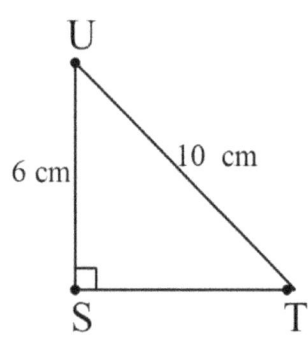

EXAMPLES

1. Find the value of x and y given that the following triangles are similar.

SOLUTION

Scale factor = $\frac{IH}{LK} = \frac{2}{5}$

To find y we multiply the size of LJ by $\frac{2}{5}$

$$\therefore y = 12 \times \frac{2}{5} = 4.8$$

To find x however, we need to multiply by the reciprocal of the scale factor $\frac{5}{2}$

$$\therefore x = 4 \times \frac{5}{2} = 10 \; cm$$

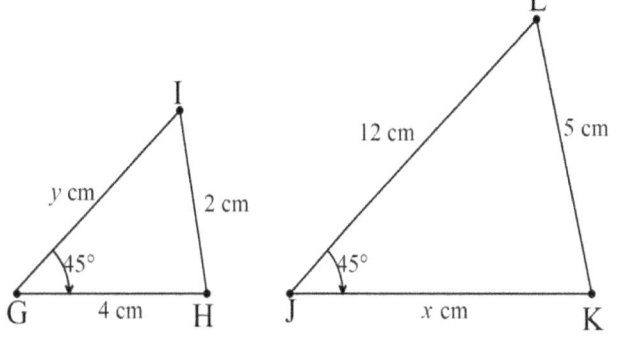

2. In the diagram BD is parallel to CE.
Given that $\triangle ABD \sim \triangle ACE$, find the value of x.

SOLUTION

Scale factor = $\frac{AB}{AC} = \frac{3}{5}$

BD corresponds to the side CE

To find x, we have to multiply 10 by $\frac{3}{5}$

$$x = 10 \times \frac{3}{5} = 6 \; cm.$$

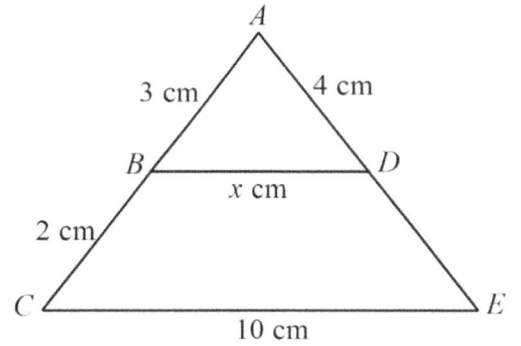

EXERCISE 10C

Use similar triangles to find the value of the pronumerals in each case.

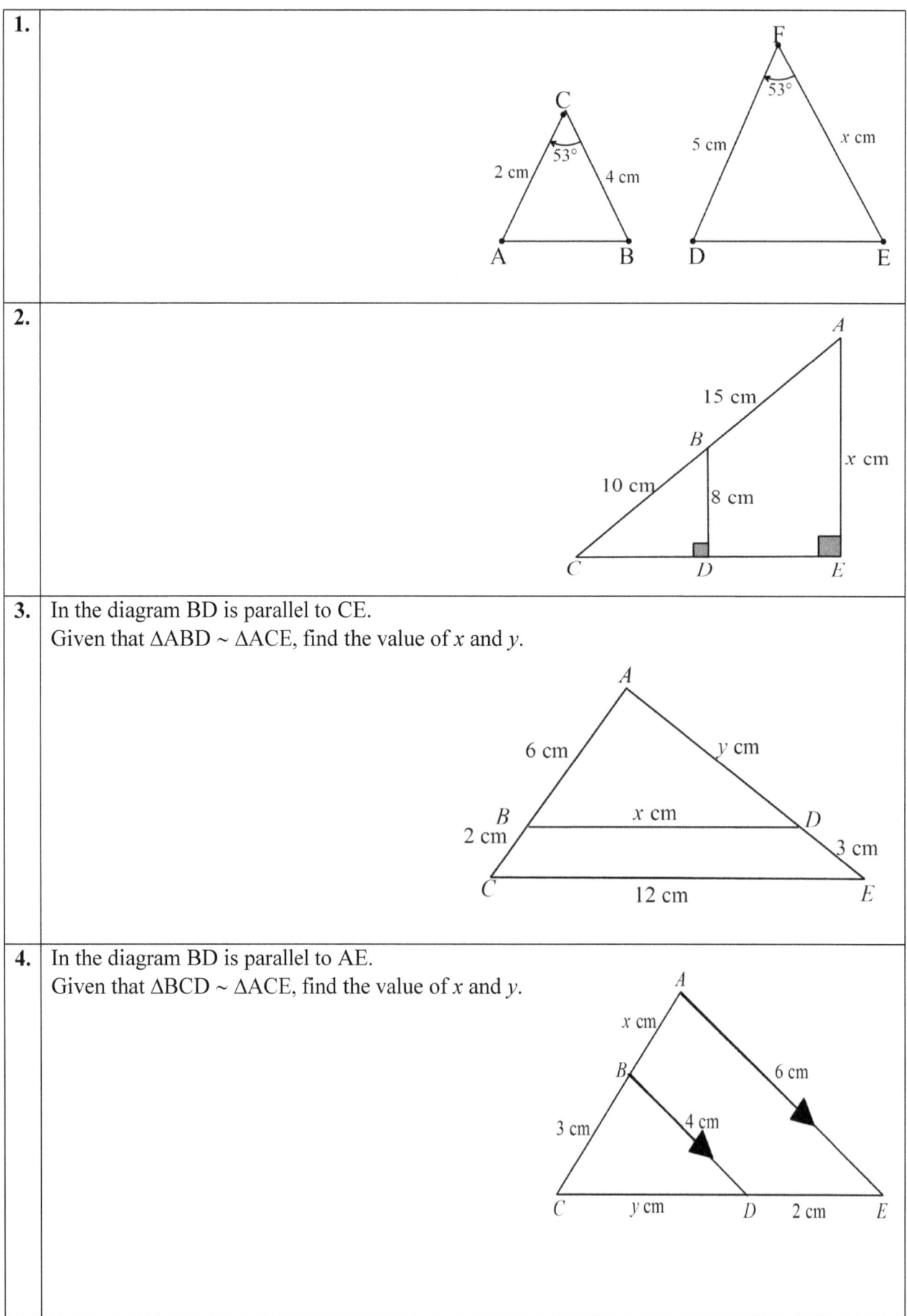

1.

2.

3. In the diagram BD is parallel to CE.
   Given that △ABD ~ △ACE, find the value of x and y.

4. In the diagram BD is parallel to AE.
   Given that △BCD ~ △ACE, find the value of x and y.

5. AOB and COD are straight lines.

   (a) Show that triangles OCA and ODB are similar.

   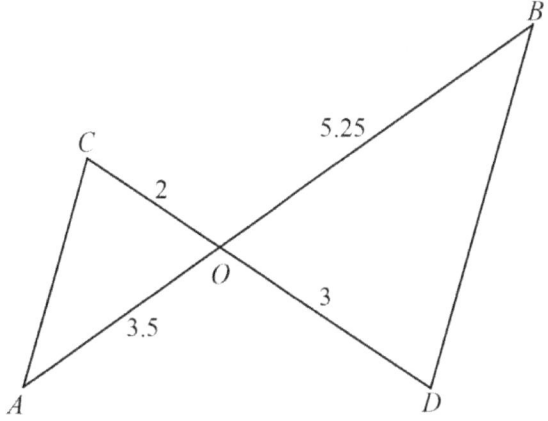

   (b) Given that BD = 3.9 cm, find AC.

6. In the given diagram, AC is parallel to DB.

   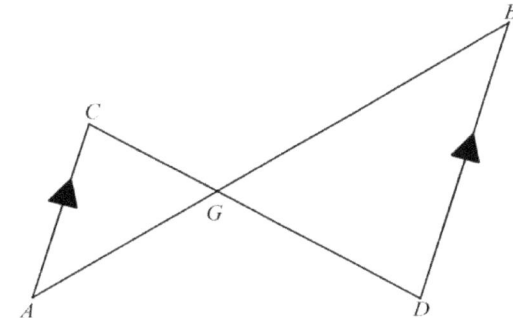

   At each step of proof below, the statement and some reasons are provided. Complete the table below.

step	Statement	Reason
1	$\angle CGA = \angle BGD$	Vertically opposite angles
2	$\angle ACG = \angle BDG$	
3	$\angle CAG = \angle DBG$	
4	$\triangle CAG$ is similar to $\triangle BDG$	AAA, SAS, SSS, RHS (choose one)

7. Show that the two triangles are similar and then find the value of *x*.

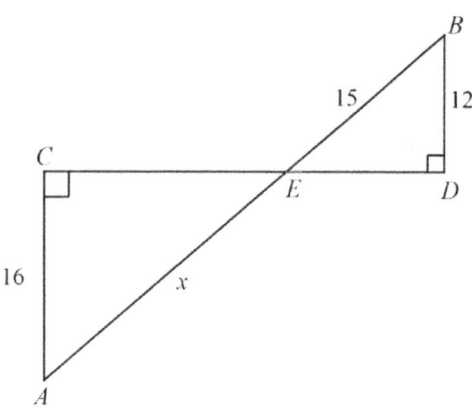

8. In the diagram, which is not drawn to scale, ∠ADB = ∠ACD = 37° and AD = DC.

(a) Write down a second pair of equal angles.

(b) Complete the two statements below:

$$\frac{AC}{AD} = \frac{\quad}{AB}$$

$$\frac{\triangle ABD}{\triangle ADC} = \frac{AD^2}{\quad}$$

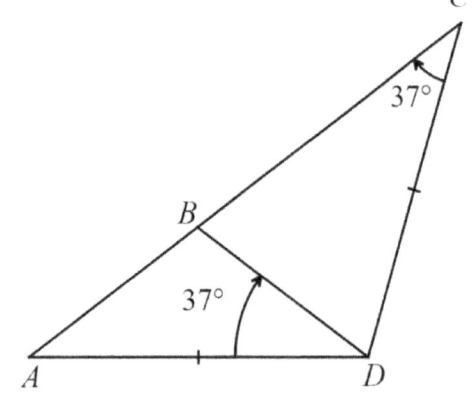

9. In the triangle ABC, AB = 6 cm, AC = 9 cm and D is the point on the side AC such that ∠ABD = ∠ACB = 40°.

(a) Write down another pair of equal angles.

(b) Use similar triangles to find the length of AD.

CHAPTER 10 : SIMILARITY

10D AREA OF SIMILAR FIGURES

If two polygons are **similar**, the ratio of their **areas** is equal to the **square** of the ratio of their corresponding sides.

In general for two similar figures having areas A_1 and A_2 and corresponding sides l_1 and l_2 respectively, $\quad \frac{A_1}{A_2} = \left(\frac{l_1}{l_2}\right)^2 \quad$ or $\quad A_1 = A_2 \times \left(\frac{l_1}{l_2}\right)^2 \quad$ or $\quad A_2 = A_1 \times \left(\frac{l_2}{l_1}\right)^2 \quad$ or $\quad \frac{l_1}{l_2} = \sqrt{\frac{A_1}{A_2}}$

EXAMPLE 1
The diagram shows two shapes. Shape 1 has an area of 12 cm². Given that the two shapes are similar, find the area of shape 2.

SOLUTION
Area of shape 2 = (Scale factor)² × Area of shape 1
$$= \left(\frac{6}{3}\right)^2 \times 12$$
$$= 48 \ cm^2$$

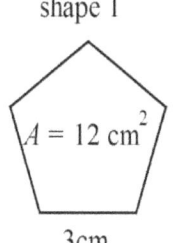
shape 1
$A = 12$ cm²
3cm

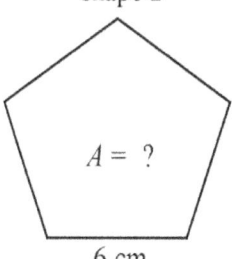
shape 2
$A = ?$
6 cm

EXAMPLE 2
The diagram shows two similar kites. The area of the larger kite is 40 cm². Find the area of the smaller kite.

SOLUTION
Area of smaller kite $= \left(\frac{3}{5}\right)^2 \times 40$
$\qquad\qquad\qquad = 14.4 \ cm^2$

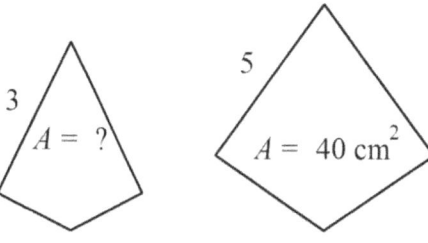

EXAMPLE 3
Two similar figures have areas 12 cm² and 27 cm² respectively. Find the ratio of their corresponding sides.

SOLUTION
Using $\frac{l_1}{l_2} = \sqrt{\frac{A_1}{A_2}}$

$$\frac{l_1}{l_2} = \sqrt{\frac{12}{27}} = \sqrt{\frac{4}{9}} = \frac{2}{3}$$

The ratio of their corresponding sides is 2 : 3.

EXAMPLE 4
The diagram shows two similar regular heptagons. Find the value of x.

SOLUTION
Using $\frac{l_1}{l_2} = \sqrt{\frac{A_1}{A_2}}$

$$\frac{3}{x} = \sqrt{\frac{27}{48}}$$
$$\frac{3}{x} = \sqrt{\frac{9}{16}} = \frac{3}{4}$$
$$\therefore x = 4$$

27 cm²
3 cm

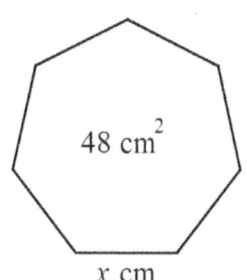
48 cm²
x cm

EXERCISE 10D

1. A photo frame and its enlargement are similar in shape. The smaller photograph has an area of 42 cm². Calculate the area of the larger photograph.

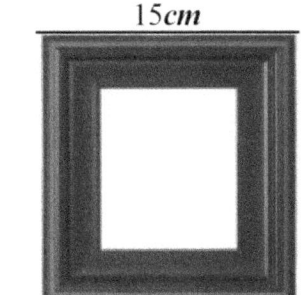

2. Two chopping boards are similar in shape. The smaller chopping board is 20 cm long and has an area of 0.03 m².
   Calculate the area of the larger chopping board.

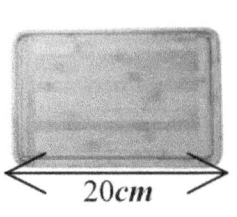 

3. Two regular heptagons are mathematically similar in shape. The larger hexagon has an area of 4600 mm².
   Find the area of the smaller hexagon.

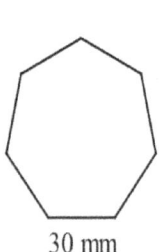 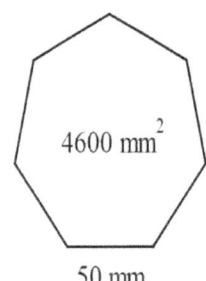

4. Two smiley faces shaped symbols are similar in shape. The smaller shape has an area of 113 cm².
   Calculate the area of the larger shape.

5. In the diagram, QR is parallel to ST and $\frac{\text{Area of } \triangle PQR}{\text{Area of } \triangle PST} = \frac{9}{64}$.

   (a) Find the value of $\frac{PQ}{PS}$,

   (b) Given that the area of the triangle PQR is 36 cm², find the area of the trapezium QRST.

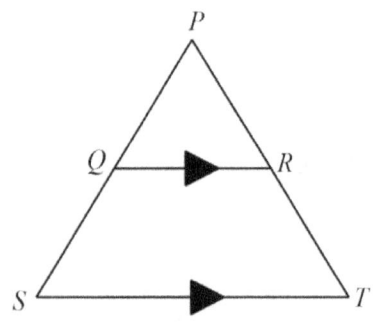

6. A rectangle has an area of 15 cm². Find the area of a similar rectangle whose dimensions are twice of the given rectangle.

7. A square has an area of 60 cm². Find the area of a square having three times the length.

8. A circle has an area of 12 cm². Find the area of a circle with radius half of the length of the given circle.

9. Find the ratio of the corresponding sides of two similar triangles if their areas are 16 cm² and 64 cm².

10. The following triangles are similar. Find the area of the triangle marked A.

24 cm², 2 cm

A, 3 cm

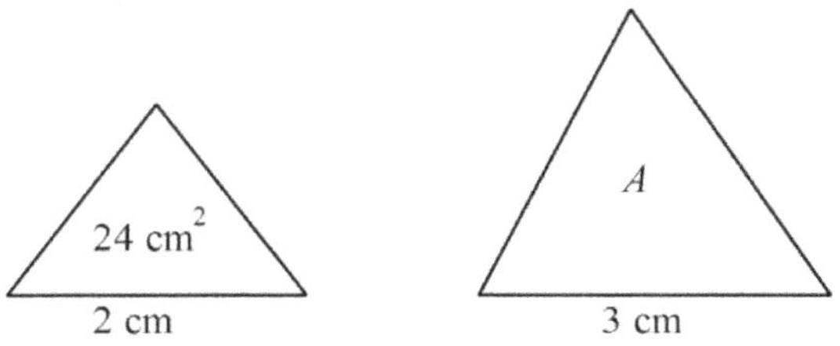

11. The following triangles are similar. Find the area of the triangle marked A.

A, 3 cm

27 cm², 5 cm

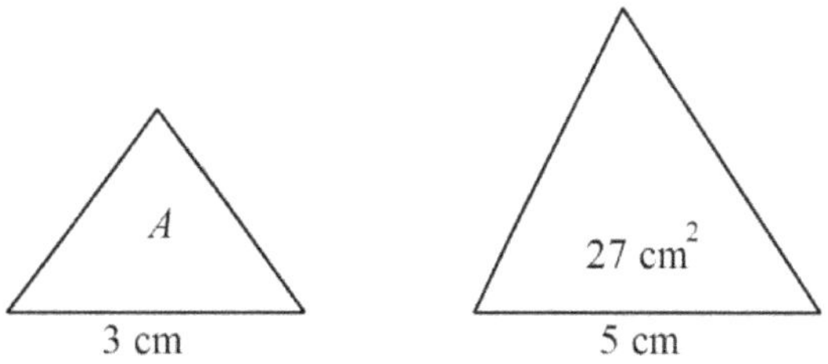

12. A floor is covered by 600 tiles which are 10 cm by 10 cm. How many 20 cm by 20 cm tiles are needed to cover the same floor?

13. A photo is 10 cm long. It is enlarged so that all dimensions are increased by 20%. Find the ratio of the area of the enlarged photo to the area of the original photo.

10E VOLUME OF SIMILAR FIGURES

If two solids are **geometrically similar**, the ratio of their **volumes** is equal to the **cube** of the ratio of their corresponding sides or radius.

In general for two similar figures having areas V_1 and V_2 and corresponding sides l_1 and l_2 respectively, $\frac{V_1}{V_2} = \left(\frac{l_1}{l_2}\right)^3$ or $V_1 = V_2 \times \left(\frac{l_1}{l_2}\right)^3$ or $V_2 = V_1 \times \left(\frac{l_2}{l_1}\right)^3$ or $\frac{l_1}{l_2} = \sqrt[3]{\frac{V_1}{V_2}}$

EXAMPLES

1. Two similar cylinders have radii 2 cm and 5 cm respectively. Find the ratio of their volumes.

   SOLUTION
   $$\frac{V_1}{V_2} = \left(\frac{r_1}{r_2}\right)^3$$
   $$= \left(\frac{2}{5}\right)^3$$
   $$= \frac{8}{125} \text{ or } 8 : 125$$

2. Two similar cones have heights 5 cm and 10 cm respectively. If the volume of the smaller cone is 25 cm³, find the volume of the larger cone.

   SOLUTION
   Scale factor $= \frac{10}{5} = 2$
   Volume of larger cone $= 2^3 \times 25$
   $= 200 \text{ } cm^3$

3. Two solid spheres have volume 80 cm³ and 270 cm³. Find the ratio of their radii.

   SOLUTION
   Using $\frac{l_1}{l_2} = \sqrt[3]{\frac{V_1}{V_2}} = \sqrt[3]{\frac{80}{270}}$
   $$= \sqrt[3]{\frac{8}{27}}$$
   $$= \frac{2}{3} \text{ or } 2 : 3$$

4. Two cylindrical containers have volumes 16 cm³ and 54 cm³ respectively. If the radius of the smaller cylinder is 6 cm, find the radius of the larger cylinder.

   SOLUTION
   Scale factor $= \sqrt[3]{\frac{V_1}{V_2}} = \sqrt[3]{\frac{54}{16}}$
   $$= \sqrt[3]{\frac{27}{8}}$$
   $$= \frac{3}{2}$$
   $radius = \frac{3}{2} \times 6 = 9 \text{ } cm$

5. Two geometrically similar jugs have volumes of 1000 cm³ and 512 cm³. They have circular bases. The diameter of the base of the larger jug is 9 cm.
   Calculate the diameter of the base of the smaller jug.

   SOLUTION
   Scale factor $= \sqrt[3]{\frac{V_1}{V_2}} = \sqrt[3]{\frac{512}{1000}} = \frac{4}{5}$

   Diameter of smaller jug $= \frac{4}{5} \times 9$
   $= 7.2 \text{ } cm.$

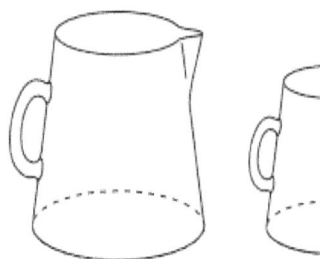

EXERCISE 10E

1. Two similar cylinders have radii 3 cm and 5 cm respectively. Find the ratio of their volumes.

2. Two spheres have radii 4 cm and 7 cm respectively, find the ratio of their volumes.

3. Two similar cylindrical jugs have volume 6 cm³ and 48 cm³. Find the ratio of their radii.

4. Two similar cylinders have heights 3 cm and 6 cm respectively. If the volume of the smaller cylinder is 36 cm³, find the volume of the larger cylinder.

5. Two similar solids have heights 2 cm and 5 cm respectively. If the volume of the larger solid is 1000 cm³, find the volume of the smaller solid.

6. Two similar jugs have volumes 200 cm³ and 25 cm³ respectively. If the radius of the smaller jug is 3 cm, find the radius of the larger jug.

7. Given that the two solids are similar, find the value of V.

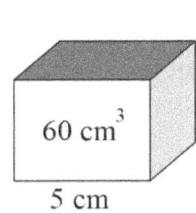 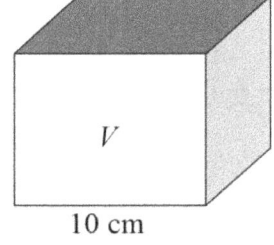

8. Given that the two cylinders are similar, find the value of V.

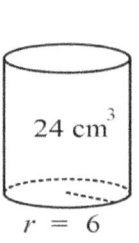 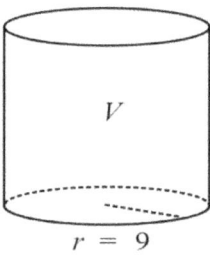

9. A cylinder K has a volume of 200 cm³. Calculate the volume of a cylinder similar to K but with radius twice that of K.

10. Mary makes two geometrically similar cakes.

    The heights of the cakes are 6 cm and 9 cm.

    Mary uses 1200 cm³ of cake mixture to make the smaller cake.
    Find the volume of cake mixture she uses to make the larger cake.

    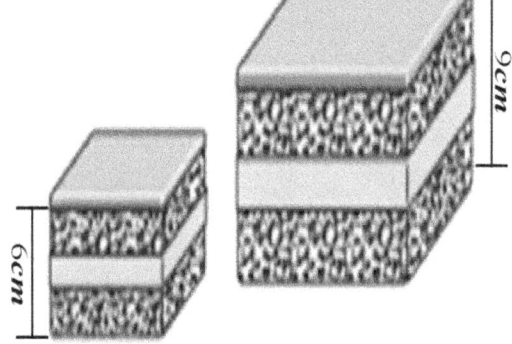

11. Two cylinders are geometrically similar. The radius of the smaller one is 3 cm. The radius of the larger one is 6 cm.

    (a) The height of the smaller cylinder is 9 cm. Find the height of the larger cylinder.

    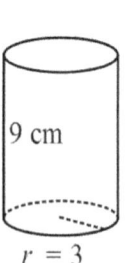
    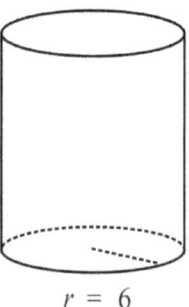

    (b) Find the ratio of the volume of the larger cylinder to the surface area of the smaller.

12. Tracy buys two tins of baked beans in Coles. The tins are geometrically similar to each other. The height of one of the tin is 8 cm and the height of the other is 12 cm.

(a) The radius of the smaller tin is 5 cm. Calculate the radius of the larger tin.

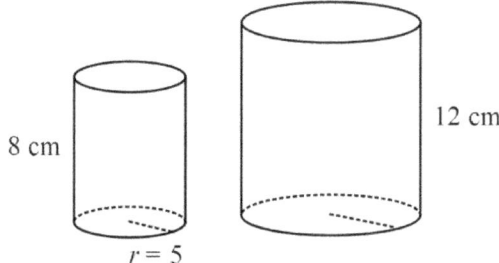

(b) Calculate the ratio of the volume of the smaller tin to the volume of the larger tin.

(c) The cost of a small tin of beans is $2.40. Calculate the cost of a large tin of beans assuming that there is no reduction for buying the larger tin.

13. The two containers shown in the diagram are geometrically similar. Their heights are 30 cm and 40 cm as shown.

(a) The radius of the smaller container is 9 cm. Calculate the radius of the base of the larger container.

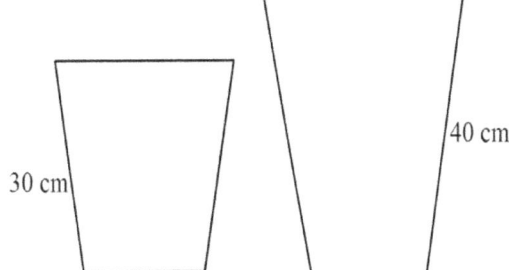

(b) The containers are completely filled with water. Given that the larger container holds 32 litres of water, calculate the capacity of the smaller container.

10 F   MAPS, SCALE DRAWINGS AND BUILDING PLANS

Maps are usually smaller than real things as it is not possible to draw a map the same size as the actual area. However, angles and directions on maps are similar to real life situations. The only difference between a map and reality is its size and we use a scale to depict this difference.

$$map\ scale = \frac{map\ length}{Actual\ length}$$

Map scales are usually written in the form 1 : n, meaning 1 cm on the map represents n cm in actual life. For example, a scale of 1 : 20 000 implies 1 cm on the map represents 20 000 cm in reality.

CLASS ACTIVITY
Write the following scales in the form 1 : n. The first 2 questions have been done as examples.

1. 1 cm represents 2 m $1\ cm \rightarrow 2\ m$ $1\ cm \rightarrow 200\ cm$ $scale\ is\ 1:200$	2. 1cm represents 3 km $1\ cm \rightarrow 3\ km$ $1\ m \rightarrow 3000 \times 100\ cm$ $scale\ is\ 1:300\ 000$	3. 5 cm represents 2 km
4. 1 cm represents 5 m	5. 1cm represents 5 km	6. 2 cm represents 8 km

7.   The diagram below shows a scale drawing of the Murray Darling Bridge in South Mississippi.

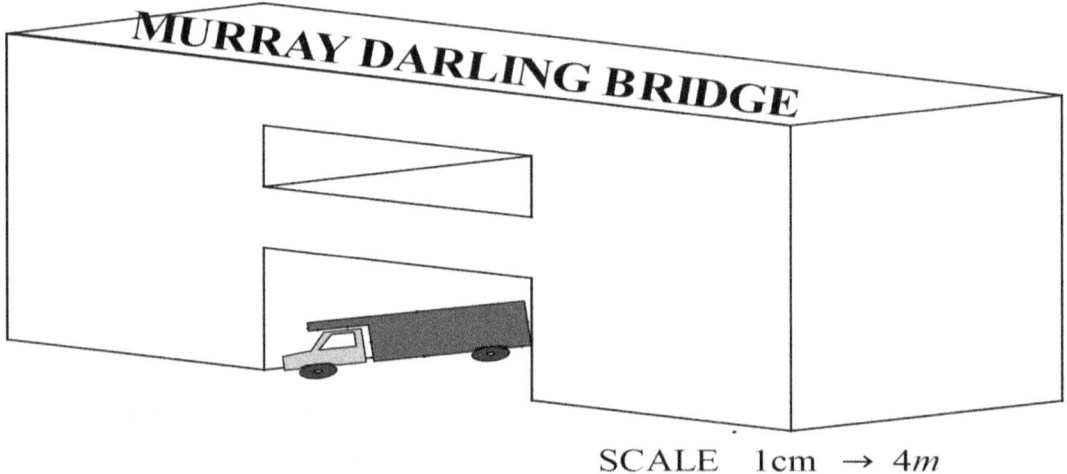

SCALE   1cm → 4*m*

By making suitable measurements, determine

(a) the height of the bridge,

(b) Santa's Christmas truck measures 4.5m high. Can it make it through the bridge?

8. The diagram shows the front elevation of Beach Side Apartments in North Carolina. Using the scale shown, determine

(a) the actual height of the building

(b) the actual width of the building.

BEACH SIDE APARTMENTS

scale
40m

9. The diagram below shows the plan of a house in Perth City.
The scale of the drawing is 1 : 100. (1cm represents 100 cm)

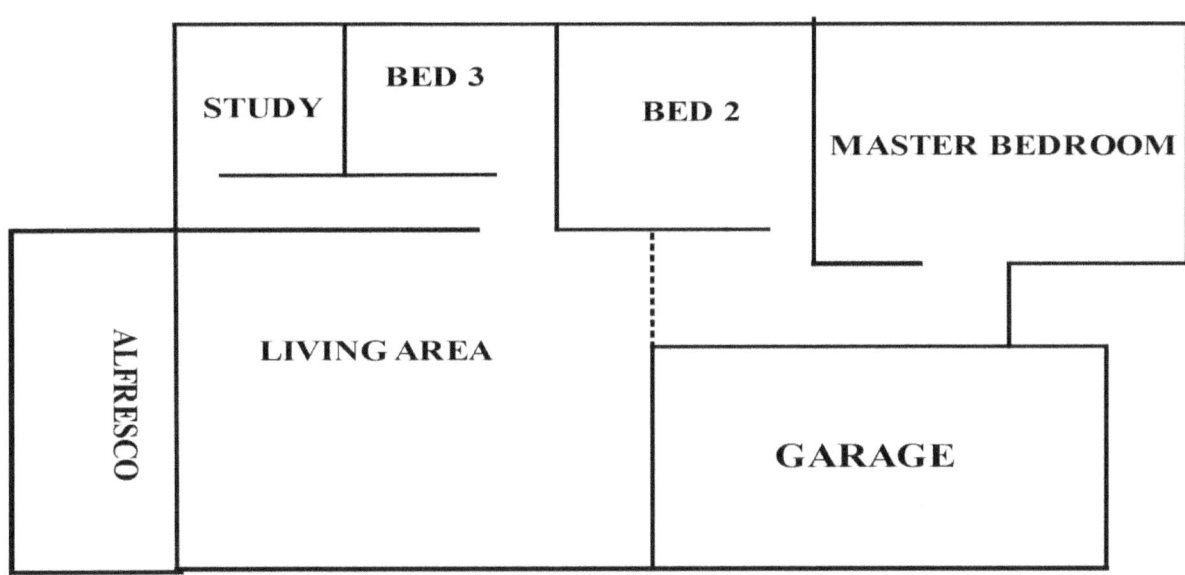

DREAM HOUSE CORPORATION

By making suitable measurements, determine

(a) The area of BED 3.

(b) The difference in area between the garage and the alfresco area.

10. The diagram shows a scale drawing of a newly found country. The scale is 1 : 50 000 000.

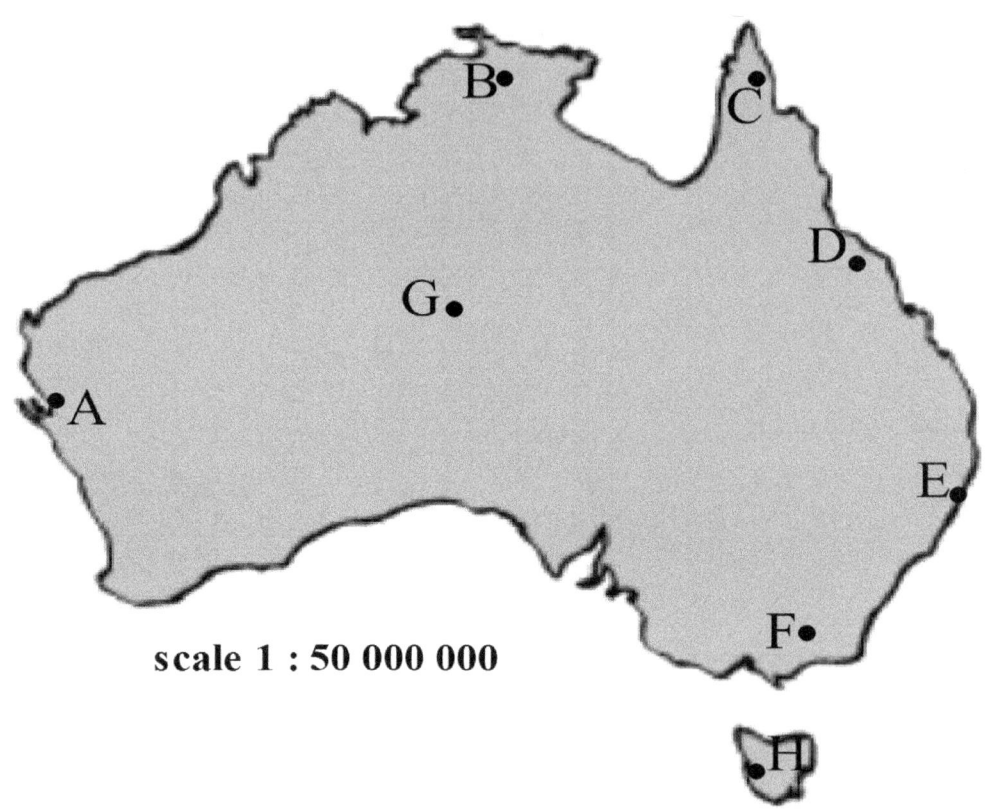

scale 1 : 50 000 000

(a) Determine how many kilometres does 1 cm on the map represent in actual.

(b) Using appropriate measurements, state the distance in kilometres between A and D.

(c) By how many kilometres is F farther from G compared to B?

MATHEMATICS APPLICATIONS UNIT 1 WORKED SOLUTIONS

CHAPTER 1

USE OF FORMULAE

1A EVALUATING LINEAR AND NON-LINEAR EXPRESSIONS

Consider the formula $T = 2\pi\sqrt{\dfrac{l}{g}}$, where T is the time period for one oscillation of a pendulum, l is the length of the pendulum and g is the acceleration due to gravity.

In mathematics, the formula $T = 2\pi\sqrt{\dfrac{l}{g}}$ is an entity constructed using the symbols. For example, to determine the time period for one oscillation (T) we need to know the length of the pendulum and the value of g in question. We can quickly and easily determine the value of T. Note that the time T and the length l are expressed as single letters instead of words or phrases.

In this chapter, the reader will be expected to substitute numerical values into algebraic expressions, and evaluate ; both linear and non-linear expressions. With complicated numerical manipulation the use of technology (CAS) is expected.

EXAMPLES

1. Given that $a = -2$ and $b = 3$, find the value of

  (i) $3a + 2b$
  (ii) $a^2 + b^2$

SOLUTION
Replace the value of a by -2 and b by 3. It is of utmost importance to always use brackets to avoid unpleasant surprises.

(i) $3(-2) + 2(3) = -6 + 6 = 0$

(ii) $a^2 + b^2 = (-2)^2 + (3)^2$
     $= 4 + 9 = 13$

2. Given that $p = 4$ and $q = -3$, evaluate

  (i) pq^2
  (ii) $(2p + q)^3$

SOLUTION

(i) $4(-3)^2 = 4 \times 9 = 36$

(ii) $[(2 \times 4 + (-3)]^3 = (8 - 3)^3$
     $= 5^3$
     $= 125$

3. Given that $a = 11$ and $b = -3$, find the value of

  (i) $\sqrt{2a - b}$
      $\sqrt{2(11) - (-3)} = \sqrt{25} = 5$

  (ii) $\sqrt[3]{2a - b + 2}$
      $\sqrt[3]{2(11) - (-3) + 2} = \sqrt[3]{27} = 3$

4. Given that $x = 3$, $y = 4$ and $z = -2$, evaluate

  (i) $2x^2 - y$
      $2(3)^2 - 4 = 18 - 4 = 14$

  (ii) $y(x - z)$
      $4(3 - (-2)) = 20$

EXERCISE 1A

1. Given that $a = 4$ and $b = -3$, find the value of

  (i) $2a + 5b$
      $2(4) + 5(-3) = -7$

  (ii) $7 - b^2$
      $7 - (-3)^2 = 7 - 9 = -2$

2. Given that $p = 3$ and $q = -5$, evaluate

  (i) $p + q^2$
      $3 + (-5)^2 = 3 + 25 = 28$

  (ii) $(3p + 2q)^2$
      $(3 \times 3 + 2 \times -5)^2 = (-1)^2 = 1$

3. Given that $a = -5$ and $b = 2$ and $c = 0$, find the value of

  (i) abc
      $-5 \times 2 \times 0 = 0$

  (ii) $a - 2b + 3c$
      $-5 - 2(2) + 3(0) = -9$

  (iii) $a^2 + b^2 - c^2$
      $(-5)^2 + 2^2 - 0^2 = 29$

4. Given that $p = -2$ and $q = 7$, evaluate

  (i) $4p + 3q$
      $4(-2) + 3(7) = -8 + 21 = 13$

  (ii) $3p^2$
      $3(-2)^2 = 3 \times 4 = 12$

  (iii) $5q - 4p$
      $5(7) - 4(-2) = 35 + 8 = 43$

5. Given that $a = 5$ and $b = -2$, find the value of

  (i) $(a - b)^2$
      $(5 - (-2))^2 = 7^2 = 49$

  (ii) $a^3 - b^3$
      $5^3 - (-2)^3 = 125 - (-8) = 133$

  (iii) a^b
      $5^{-2} = \dfrac{1}{25}$

6. Given that $p = 2$, $q = -3$ and $r = 4$, evaluate

  (i) $4p + q - 2r$
      $4(2) + (-3) - 2(4) = -3$

  (ii) $5p^2$
      $5(2)^2 = 5 \times 4 = 20$

  (iii) $(p + 2r)(q + 2r)$
      $(2 + 2 \times 4)(-3 + 2 \times 4)$
      $= (2 + 8)(-3 + 8)$
      $= 10 \times 5 = 50$

CHAPTER 1 : USE OF FORMULAE SOLUTIONS

1B DETERMINING THE VALUE OF THE SUBJECT OF A FORMULA

At this stage of the chapter, we have to determine the value of the subject of a formula given the values of the other pronumerals in the formula. (Transposition not required)

EXAMPLE 1

Given the formula $y = mx + c$, find y when $m = 4, x = -2$ and $c = 11$.
Solution
$y = 4(-2) + 11 = 3$

EXAMPLE 2

Given that $v^2 = u^2 + 2as$, find the values of v when $u = 3, a = 10$ and $s = 3.6$.
Solution
$v^2 = 3^2 + 2(10)(3.6) = 81$
$\therefore v = \pm 9$

EXERCISE 1B

1. Given the formula $v = u + at$, find v when $u = 25, a = 10$ and $t = 3$.

    $v = 25 + 10(3) = \mathbf{55}$

2. Given the formula $s = ut + \frac{1}{2}at^2$, find the value of s when $u = 20, a = 9.8$ and $t = 6$.

    $s = 20(6) + \frac{1}{2}(9.8)(6)^2 = \mathbf{296.4}$

3. Given the formula $A = 2\pi r^2 + 2\pi rh$, find A when $r = 4$ and $h = 9$. (Use $\pi = 3.14$)

    $A = 2 \times 3.14 \times 4^2 + 2 \times 3.14 \times 4 \times 9$
    $= \mathbf{326.56}$

4. Given that $V = \frac{4}{3}\pi r^3$, find the value of V when $r = 5$.

    $V = \frac{4}{3} \times \pi \times 5^3 = \mathbf{523.60}$

5. The sum (S) of positive integers from 1 to n is given by $S = \frac{1}{2}n(n+1)$. Find S when $n = 12$.

    $S = \frac{1}{2} \times 12(12+1) = \mathbf{78}$

6. Given that the area (A) of a triangle is given by $A = \frac{ab\sin C}{2}$, find the value of A when $a = 5, b = 10$ and $C = 30°$.

    $A = \dfrac{5 \times 10 \times \sin 30}{2} = \mathbf{12.5}$

7. Einstein's famous equation relating energy (E), mass (m) and speed of light (c) is $E = mc^2$.
Find E when $m = 0.0001$ and $c = 3 \times 10^8$.

    $E = 0.0001 \times (3 \times 10^8)^2 = \mathbf{9 \times 10^{12}}$

8. Given that $y = a + bx^2$, find the value of y when $a = 7, b = 4$ and $x = -5$.

    $y = 7 + 4(-5)^2 = 7 + 4 \times 25 = \mathbf{107}$

1C USE OF FORMULAE AND TECHNOLOGY (CAS)

Where the numerical manipulation is complicated, the use of technology is really helpful. Use the following steps on your CAS:

- Menu
- Num Solve
- Insert your formula
- Input given values
- Solve

EXAMPLES

Use the solve facility on your calculator to attempt the following questions.

1. Given that $v^2 = u^2 + 2as$, find the value of v ($v > 0$) when $u = 5, a = 6$ and $s = 50$.

Remember to click the bubble at v as we are solving for v.

$v = 25$

Equation:
$v^2 = u^2 + 2 \times a \times s$

● $v =$
○ $u = 5$
○ $a = 6$
○ $s = 50$

Lower = -9E+999
Upper = 9E+999

2. Given that $c = b^2 \times (a+b) + b^3 - a^2$, find c when $a = 3$ and $b = 5$.

SOLUTION

$c = 316$

Equation:
$c = b^2 \times (a+b) + b^3 - a^2$

● $c =$
○ $a = 3$
○ $b = 5$

Lower = -9E+999
Upper = 9E+999

CHAPTER 1 : USE OF FORMULAE SOLUTIONS

EXERCISE 1C

Use Num Solve to find the missing pronumerals in each case.

1. $V = \frac{1}{3}\pi r^2 h$

Evaluate V

(a) when $r = 6$ and $h = 12$.

452.39

(b) when $r = 5$ and $h = 10$.

261.80

(c) when $r = 8$ and $h = 52$.

3485.07

2. $s = \frac{1}{2}(u+v)t$, find the value of s

(a) when $u = 30, v = 50$ and $t = 2$.

80

(b) when $u = 40, v = 60$ and $t = 3$.

150

(c) when $u = 20, v = 48$ and $t = 3$.

102

3. The formula for the area of a trapezium is $A = \frac{1}{2}h(c+d)$.

Evaluate A when

(a) $h = 8, c = 11$ and $d = 13$.

96

(b) $h = 16, c = 15$ and $d = 25$.

320

4. Find the value of $A = \frac{a+\sqrt{a^2+b^2}}{a^2-2ab}$

when $a = -4$ and $b = -3$. Give your answer as a fraction.

$-\frac{1}{8}$

5. $\frac{1}{b} = \frac{1}{c} + \frac{1}{d}$,

Evaluate b when $c = 3$ and $d = 8$.

2.18

6. $\frac{1}{x} = \frac{1}{y} + \frac{1}{z}$,

Evaluate x when $y = 4$ and $z = 10$.

$\frac{20}{7}$

7. Given $v = u + at$, find v given that

(a) $u = 10, a = 5$ and $t = 7$.

45

(b) $u = 55, a = 10$ and $t = 4$.

95

8. If $E = \frac{1}{2}mv^2$, find the value of E when

(a) $m = 100$ and $v = 5$.

1250

(b) $m = 200$ and $v = 25$.

62500

9. In the formula $y = mx + c$, find y given that

(a) $m = 5, x = 10$ and $c = 3$.

53

(b) $m = -5, x = 8$ and $c = 11$.

−29

(c) $m = 0.5, x = 12$ and $c = 17$.

23

10. In the formula $E = mc^2$, find E given that

(a) $m = 50$ and $c = 5$.

1250

(b) $m = 60$ and $c = 8$.

3840

11. Given that $v^2 = u^2 + 2as$, find the value of v when

(a) $u = 20, a = 8$ and $s = 50$.

± 34.64

(b) $u = 30, a = 10$ and $s = 100$.

± 53.85

12. Given that $A = \frac{a+b}{2} \times h$, find the value of A when $a = 18, b = 12$ and $h = 9$.

135

13. $V = \pi r^2 h$, find the value of V given that

(a) $r = 8$ and $h = 12$.

2412.74

(b) $r = 7$ and $h = 20$.

3078.76

14. $V = 2\pi r^2 + 2\pi rh$, find the value of V given that $r = 6$ and $h = 11$.

640.88

1D APPLICATIONS

1. Body Mass Index (BMI) is an index of weight-for-height that is commonly used to classify underweight, overweight and obesity in adults. It is defined as the weight (W) in kilograms divided by the square of the height (H) in metres (kg/m²). Thus,

$$BMI = \frac{W}{H^2}$$

(a) Use the above formula to determine the BMI of an adult who weighs 70kg and whose height is 1.75m.

$$BMI = \frac{70}{1.75^2} = 22.86$$

(b) A BMI of 25 or more is classified as over-weight. Sarah weighs 80kg and has a height of 1.65m. Is she overweight? Support your answer by calculations.

$$BMI = \frac{80}{1.65^2} = 29.4,$$

Sarah is well overweight.

(c) A BMI of less than 18.5 is normally considered as underweight. Peppa is 51kg and has a height of 150 cm. Calculate Peppa's BMI to see whether he is under weight or not.

$$BMI = \frac{51}{1.5^2} = 22.7,$$

∴ *Peppa is not underweight.*

2. The formula used to convert Fahrenheit (F) to Celsius (C) is given by $C = \frac{5}{9}(F - 32)$.
Use the formula to convert

(a) 48°F into °C.

$$\frac{5}{9}(48 - 32) = 8.9°C$$

(b) 108°F into °C.

$$\frac{5}{9}(108 - 32) = 42.22°C$$

(c) 49°C into °F.

$$\frac{5}{9}(F - 32) = 49°C \therefore F = 120.2$$

3. The Kinetic Energy (E) is the energy possessed by a body because of its motion and is equal to half the mass (m) of the body times the square of its speed (v). Symbolically, $E = \frac{1}{2}mv^2$.

(a) Calculate the kinetic energy of a particle having mass 50kg and speed 20 m/s.

$$E = \frac{1}{2} \times 50 \times 20^2 = 10000\,J$$

(b) Calculate the kinetic energy of a car which has a mass of 1000 kg and is moving at the rate of 25 m/s.

$$E = \frac{1}{2} \times 1000 \times 25^2 = 312\,500\,J$$

(c) What is the speed of a horse weighing 345 kg and having a kinetic energy of $1.725 \times 10^4 J$?

$$1.725 \times 10^4 = \frac{1}{2} \times 345 \times v^2$$

$$v = 10\,m/s$$

4. The formula that can be used to find the compound interest is given by

$$A = P\left(1 + \frac{R}{n \times 100}\right)^{n \times T}$$

where A is the principal plus interest, P means Principal value, R is the rate of interest, T is time in years and n is the number of times interest is compounded per year.

(a) Amy invests $3500 in a bank for 5 years at 5.48% per annum, where the interest is compounded quarterly. Find the value of her investment at the end of the 5 years.

$$A = 3500\left(1 + \frac{5.48}{4 \times 100}\right)^{4 \times 5} = \$4594.70$$

(b) Paul borrows $10800 for 4 years at 7.2% per annum, where the interest is compounded monthly. Find the interest paid.

$$A = 10800\left(1 + \frac{7.2}{12 \times 100}\right)^{12 \times 4} = \$14392.19$$

$$Interest = 14392.19 - 10800 = \$3592.19$$

CHAPTER 1 : USE OF FORMULAE SOLUTIONS

5. Impulse (I) is defined as the change in the momentum of a body caused over a very short time. If m is the mass and v and u the final and initial velocities of a body, then
$I = m(v - u)$.

   (a) Calculate the impulse of a body of mass 5 kg whose speed increases from 10 m/s to 15 m/s in a short amount of time.

   $$I = 5(15 - 10) = 25 \, Ns$$

   (b) Calculate the mass of a body having an impulse of 480 Ns when its speed increases from 18 m/s to 30 m/s.

   $$480 = m(30 - 18)$$
   $$\therefore m = 40 \, kg$$

6. Elastic potential energy (PE) is defined as the Potential energy stored as a result of deformation of an elastic object. It is equal to the work done to stretch the spring, which depends upon the spring constant k as well as the distance stretched (L).
The formula used to calculate PE is given by
$PE = \frac{1}{2}kL^2$.

   (a) Determine how much Elastic Potential energy a spring with a spring constant of 12 N/m stores if it is stretched by 1.4m.

   $$PE = \frac{1}{2} \times 12 \times 1.4^2 = 11.76 \, J$$

   (b) Determine the spring constant k for a spring having elastic Potential energy 27.04 J and stretched a distance of 2.6m.

   $$27.04 = \frac{1}{2} \times k \times 2.6^2$$
   $$\therefore k = 8 \, N/m$$

MATHEMATICS APPLICATIONS UNIT 1

7. The period of a pendulum (T) can be worked out using the formula
$$T = 2\pi\sqrt{\frac{l}{g}}$$
where T is the time in seconds, l is the length of the pendulum (in metres) and g is the acceleration due to gravity in m/s² (use $g = 10$ m/s²)

   (a) Use this formula to determine the period of a pendulum having length 0.60 m.

   $$T = 2\pi\sqrt{\frac{0.60}{10}} = 1.54 \, s$$

   (b) What would be the period of a pendulum if it is 1.5m long?

   $$T = 2\pi\sqrt{\frac{1.5}{10}} = 2.43 \, s$$

   (c) If the pendulum's length in (b) were to be shortened by one-third its original value, what would be its new period?

   $$New \, length = 1m$$
   $$New \, period \, T = 2\pi\sqrt{\frac{1}{10}} = 1.99 \, s$$

8. If m_1 and m_2 are the masses of two different objects and d is the distance between the centres of the two objects, then the gravitational attraction (F) between the two objects also known as Newton's Law of Gravity is given by $F = \frac{Gm_1m_2}{d^2}$ where G is the gravitational constant (Use $G = 6.7 \times 10^{-11}$).

   (a) Two objects have masses 200 000 kg and 300 000 kg respectively. The centres of the two objects are 0.5m apart. Determine the gravitational attraction between the two objects.

   $$F = \frac{6.7 \times 10^{-11} \times 200000 \times 300000}{0.5^2} = 16.08 \, N$$

   (b) Explain what happens to the value of F as the distance between the objects decreases.

   The value of F increases

1E SPREADSHEETS

The objective of using spreadsheets is to learn how to use formulae to perform calculations and understand how we use cell addresses within a formula.

There are some important rules governing spreadsheets which we have to apply when solving problems.

- ❖ All formulae start with an = sign to identify them as a formula.
- ❖ We use the asterix (*) for multiplication
- ❖ For division make use of the forward slash (/)

EXAMPLE 1

We are going to use the formula $y = mx + c$ and a spreadsheet to calculate values of y for different sets of values of m, x and c.

Consider the second row, to calculate y enter = $A2*B2 + C2$ in cell D2 as shown and the answer will appear as 14.
Similarly in cell D3 enter =$A3*B3 + C3$ and the value of 22 will automatically come up in that cell.
To fill the rest of the spreadsheet we can just drag and drop and the other cells D4, D5 etc.. will be worked out automatically.

	A	B	C	D
1	m	x	c	y
2	2	5	4	= A2*B2 + C2 14
3	3	4	10	= A3*B3 + C3 22
4	5	3	-7	8
5	10	-2	-5	-25
6	8	4	6	38

EXAMPLE 2

The spreadsheet below shows the T-Shirt sales of Top Clothing Ltd. Insert appropriate formula in cells G2, G3 and so on to find the total number of shirts sold for each size.
Consider the second row, to calculate the total enter = **(sum B2 : F2)** in cell D2 as shown and the answer 76 will be displayed. The rest of column G can as usual be dragged and filled.

	A	B	C	D	E	F	G
1		Monday	Tuesday	Wednesday	Thursday	Friday	Total
2	X Small	12	15	10	21	18	=(sum B2 : F2) 76
3	Small	10	11	8	13	12	54
4	Medium	25	21	33	41	38	158
5	Large	8	7	5	10	6	36

EXERCISE 1E

1. Use a spreadsheet from your calculator to complete the following.

	A	B	C	D
1	Quantity sold	Price	Sales	GST
2	8	5	= A2*B2 40	= 0.1*C2 4
3	10	9	90	9
4	15	4	60	6
5	7	20	140	14
6	22	3	66	6.60

2. The spreadsheet below shows the marks of a group of students studying Application Mathematics in four different tests. The tests were marked out of 10 marks each. Use appropriate formula in your spreadsheet to complete columns F and G.

	A	B	C	D	E	F	G
1		Test 1	Test 2	Test 3	Test 4	Total Marks	%
2	Alice	9	7	5	8	=(sum B2 : F2) = 29	= 2.5*F2 = 72.5
3	Berry	5	4	6	4	19	47.5
4	Carol	9	9	8	9	35	87.5
5	Damien	10	8	8	5	31	77.5
6	Essen	2	4	5	1	12	30

3. The area of a trapezium is $A = \frac{a+b}{2} \times h$. Use the spreadsheet below and your calculator to complete the table for different sets of values of a, b and h.

	A	B	C	D
1	a	b	h	Area
2	8	6	10	$= 0.5(A2 + B2) * C2$ 70
3	11	19	5	75
4	15	17	8	128
5	7	23	20	300
6	19	21	8	160

CHAPTER 2

PERCENTAGES

A percentage is indicated by the symbol %, and it means a fraction of one hundred.

Example $13\% = \frac{13}{100} = 0.13$

It is customary to express a fraction in its simplest form.

Example $15\% = \frac{15}{100} = \frac{3}{20}$

CLASS ACTIVITY

Complete the following table

Percentage	Fraction	Decimal	Percentage	Fraction	Decimal
17%	$\frac{17}{100}$	0.17	60%	$\frac{60}{100} = \frac{3}{5}$	0.6
30%	$\frac{30}{100} = \frac{3}{10}$	**0.3**	**48%**	$\frac{12}{25}$	**0.48**
31%	$\frac{31}{100}$	**0.31**	65%	$\frac{13}{20}$	0.65
40%	$\frac{2}{5}$	0.4	**64%**	$\frac{16}{25}$	**0.64**
36%	$\frac{9}{25}$	**0.36**	**26%**	$\frac{13}{50}$	**0.26**
210%	$2\frac{10}{100} = 2\frac{1}{10}$	2.1	325%	$3\frac{25}{100} = 3\frac{1}{4}$	**3.25**
145%	$1\frac{45}{100} = 1\frac{9}{20}$	1.45	**420%**	$4\frac{1}{5}$	**4.2**

2A EXPRESSING ONE QUANTITY AS A PERCENTAGE OF ANOTHER

A percentage can be used to compare two quantities expressed in the same units.
Consider a Year 11 class of 24 students studying Mathematics Applications.
If 16 students live in the same suburb, we can express the students living in the same suburb as a percentage of the whole class.

Thus fraction of students living in the same suburb = $\frac{Number\ of\ students\ living\ in\ the\ same\ suburb}{total\ number\ of\ students\ in\ the\ class}$

To convert fraction into percentage, we multiply by 100 as usual.

Percentage of students living in the same suburb = $\frac{16}{24} \times 100 = 66.7\ \%$

EXAMPLE

1. What percentage is 32 out of 50?

SOLUTION

$\frac{32}{50} \times 100 = 64\ \%$

2. Express 400 metres as a percentage of 2.45 km.

SOLUTION

2.45 km = 2450 m

$\frac{400}{2450} \times 100 = 16.33\ \%$

3. Express 93g as a percentage of 3 kg.

SOLUTION

3 kg = 3000 g

$\frac{93}{3000} \times 100 = 3.1\ \%$

4. Express 2500 m² as a percentage of 12 hectares (1 ha = 10 000m²).

SOLUTION

12 ha = 120000 m²

$\frac{2500}{120000} \times 100 = 2.1\ \%$

5. An airline company has seats for 120 passengers. Calculate

(a) The number of passengers on board when $\frac{7}{15}$ of the seats are occupied.

$\frac{7}{15} \times 120 = 56$

(b) The percentage of seats which are occupied when there are 114 passengers on board.

$\frac{114}{120} \times 100 = 95\%$

CHAPTER 2 : PERCENTAGES SOLUTIONS

EXERCISE 2A

1. What percentage is 11 out of 20?

    $$\frac{11}{20} \times 100 = 55\,\%$$

2. Express 400 metres as a percentage of 3.6 km.

    $$\frac{400}{3600} \times 100 = 11.1\,\%$$

3. Express 120g as a percentage of 5 kg.

    $$\frac{120}{5000} \times 100 = 2.4\,\%$$

4. Express 36000 m² as a percentage of 15 hectares.

    $$\frac{36000}{150000} \times 100 = 24\,\%$$

5. A theatre has seats for 1100 people. Calculate the percentage of seats that are occupied if 847 people attend.

    $$\frac{847}{1100} \times 100 = 77\,\%$$

6. To make a chicken and mushroom pie, Mrs Mac requires 550g of chicken and 125g of mushroom. Express the mass of mushroom as a percentage of the mass of the chicken, giving your answer correct to the nearest whole number.

    $$\frac{125}{550} \times 100 = 22.73\,\%$$

7. A market gardener sows 25000 lettuce seeds. Given that 21500 produce seedlings, calculate the percentage which did **not** produce seedlings.

    $$25000 - 21500 = 3500$$

    $$\frac{3500}{25000} \times 100 = 14\,\%$$

8. In September 2012 the number of students at a school was 540. Given that the number of students at the school in September 2013 was 567, find the percentage increase in the number of students.

    $$567 - 540 = 27$$

    $$\frac{27}{540} \times 100 = 5\,\%$$

9. Jessy earns $1500 each month. He pays rent of $525 each month. Find the amount he pays in rent as a percentage of his earnings.

    $$\frac{525}{1500} \times 100 = 35\,\%$$

10. In a sale, the price of bicycle B is reduced from $2400 to $1596. Calculate the percentage reduction given.

    $$2400 - 1596 = 804$$

    $$\frac{804}{2400} \times 100 = 33.5\,\%$$

2B FINDING A GIVEN PERCENTAGE OF AN AMOUNT

To find a given percentage of an amount, use the following steps:

- Write the given percentage as a fraction
- Multiply the amount by the fraction

EXAMPLE

1. Find 12% of 300.

    Solution

    $$12\% \text{ of } 300 = \frac{12}{100} \times 300 = 36$$

2. Find 15% of $10.80.

    Solution

    $$15\% \text{ of } \$10.8 = \frac{15}{100} \times 10.80$$
    $$= \$1.62$$

EXERCISE 2B

1. Find 13% of 400.

    $$\frac{13}{100} \times 400 = 52$$

2. Find 8% of $25.60.

    $$\frac{8}{100} \times 25.60 = \$2.05$$

3. Find 125% of 200 kg.

    $$\frac{125}{100} \times 200 = 250\,kg$$

4. Find 24% of 90 litres.

    $$\frac{24}{100} \times 90 = 21.6\,litres$$

5. Find 21% of $800.

    $$\frac{21}{100} \times 800 = \$168$$

6. Find 20% of $500

    $$\frac{20}{100} \times 500 = \$100$$

7. On a packet of tomato seeds it is stated that 93% should produce seedlings. Find the number of seedlings which would be expected from a packet containing 600 seeds.

    $$\frac{93}{100} \times 600 = 558$$

8. Of a class of 60 students, 25% play soccer, 35% play tennis and the rest play netball. Calculate the number of students who play netball.

    $$100 - (25 + 35) = 40$$

    $$\frac{40}{100} \times 60 = 24$$

CHAPTER 2 : PERCENTAGES SOLUTIONS

2C INCREASE AND DECREASE IN PERCENTAGES

To increase or decrease an amount by a certain percentage, it is easier to multiply the amount by a multiplying factor as shown below.

CLASS ACTIVITY

Complete the table.

Increase by	Multiplying factor	Decrease by	Multiplying factor
8%	100% + 8% = 108% = 1.08	12%	100% − 12% = 88% = 0.88
15%	1.15	2.5%	0.975%
7%	1.07	4%	0.96%
13%	1.13	15%	0.85%
125%	2.25	90%	0.1
4.5%	1.045	4.5%	0.955%
8.5%	1.085	0.5%	0.995%

EXAMPLE

1. Increase 80 by 10%.
   SOLUTION
   $$80 \times 1.10 = 88$$

2. Decrease 200 by 5%.
   SOLUTION
   $$200 \times 0.95 = 190$$

3. Due to an increase in the cost of living, Tom's salary increased from $16.80 per hour by 3%. Calculate Tom's new hourly rate.
   SOLUTION
   $$16.80 \times 1.03 = \$17.30$$

4. In 2013 the number of tourists visiting Monkey Mia was 15600. This number decreases by 2.5 % in 2014. Find the number of tourists who visited Monkey Mia in 2014.
   SOLUTION
   $$100 - 2.5 = 97.5\% = 0.975$$
   $$15600 \times 0.975 = 15210$$

EXERCISE 2C

1. Increase 300 by 11%.
   $$300 \times 1.11 = 333$$

2. Decrease $400 by 6%.
   $$400 \times 0.94 = 376$$

3. Increase 500 kg by 15%.
   $$500 \times 1.15 = 575\ kg$$

4. Decrease 4600 m by 7%.
   $$4600 \times 0.93 = 4278\ m$$

5. Increase 1600 km by 9%.
   $$1600 \times 1.09 = 1744\ km$$

6. Decrease $1600 by 3.5%.
   $$1600 \times 0.965 = \$1544$$

7. Increase 55 kg by 2.3%.
   $$55 \times 1.023 = 56.265\ kg$$

8. Decrease 450m by 23%.
   $$450 \times 0.77 = 346.5\ m$$

9. A clerk's monthly salary is $5600. If his monthly salary increases by 4.2%, calculate his new monthly salary.
   $$5600 \times 1.042 = \$5835.20$$

10. In 2014, the number of tourists visiting Perth increased by 5% compared to 124000 in 2013. Calculate the number of tourists visiting Perth in 2014.
    $$124000 \times 1.05 = 130200$$

11. A factory produces 15000 toys each year. Due to a slowdown in the economy, the manager decides to reduce the production capacity by 7.5%. Calculate the new production level.
    $$15000 \times 0.925 = 13875$$

12. A man earning $5625 is awarded a pay rise of 8%. Calculate his new salary wage.
    $$5625 \times 1.08 = \$6075$$

13. The price of X-Box 360 decreases from $360 to $280. What is the percentage decrease in price?
    $$\frac{80}{360} \times 100 = 22.2\%$$

14. In 2013, the cost of posting a letter was increased from 60 cents to 72 cents. Calculate the percentage increase.
    $$\frac{12}{60} \times 100 = 20\%$$

CHAPTER 2 : PERCENTAGES SOLUTIONS

2D CALCULATING AN ORIGINAL AMOUNT

EXAMPLES

1. 15% of an amount is 40, find the amount.
SOLUTION
Amount $= 40 \times \frac{100}{15} = 266.67$

2. 5% of a ticket cost $2.40. Find the cost of the ticket.
SOLUTION
Cost of ticket $= 2.40 \times \frac{100}{5} = 48$

3. Hotel expenses, totalling $234, accounts for 65% of the cost of the holiday. Calculate the total cost of the holiday.
SOLUTION
Cost of holiday $= 234 \times \frac{100}{65} = \$360$

4. There were 756 children in a school. This is 5% more than it was last year. Calculate the number of children in the school last year.
SOLUTION
Number of children $= 756 \times \frac{100}{105} = 720$

EXERCISE 2D

1. 12% of an amount is 66, find the amount.

$$66 \times \frac{100}{12} = 550$$

2. 8% of a cinema ticket cost $1.92. Find the cost of the ticket.

$$1.92 \times \frac{100}{8} = \$24$$

3. 35% of an amount is 210, find the amount.

$$210 \times \frac{100}{35} = 600$$

4. 25% of a number is 72. Find the number.

$$72 \times \frac{100}{25} = 288$$

5. If 121 students went on holiday last year, accounting for 11% of a school population, find the number of students at the school.

$$121 \times \frac{100}{11} = 1100$$

6. There were 903 children in a school. This is 5% more than it was last year. Calculate the number of children in the school last year.

$$903 \times \frac{100}{105} = 860$$

7. Tax on the original price of a bicycle is charged at 20% of the original price. After tax has been included, Matthew pays $1080 for this bicycle. Calculate the original price.

$$1080 \times \frac{100}{120} = \$900$$

8. In June 2012, there were 450 members at a tennis club. Given that this was 20% more than that in June 2011, find the number of members in June 2011.

$$450 \times \frac{100}{120} = 375$$

2E PROFIT AND LOSS

A businessman makes a profit on selling an article if the selling price is more than the cost price. On the other hand, a businessman who sells an article at a price lesser than the cost price incurs a loss.

Hence Profit = Selling Price − Cost Price
  Loss = Cost Price − Selling Price

EXAMPLES

1. A trader bought a calculator for $60 and sold it for $75. Calculate the percentage profit.
SOLUTION
Profit $= \$75 - \$60 = \$15$
Percentage profit $= \frac{profit}{cost\ price} \times 100$
$= \frac{15}{60} \times 100 = 25\%$

2. Alex bought a DVD player for $90 and sold it for $80. Calculate his percentage loss.
SOLUTION
Loss $= \$90 - \$80 = \$10$
Percentage loss $= \frac{loss}{cost\ price} \times 100$
$= \frac{10}{90} \times 100 = 11.1\%$

3. For her formal, Ashley bought a dress on EBay for $400 and sold it a year later at a profit of 8%. Calculate the selling price of the dress.
SOLUTION
Selling price $= 1.08 \times 400 = \$432$

4. JR Trading sold a toy for $76.50 incurring a loss of 15%. Find the cost price of the toy.
SOLUTION
Cost price $= 76.50 \times \frac{100}{85} = \$90$

5. During the end of year financial sale, all prices are reduced by 40%. Calculate the original price of an article whose sale price is $48.
SOLUTION
Original price $= 48 \times \frac{100}{60} = \$80$

6. When Tony, the salesman, sells a camera for $84 he makes a profit of 12%. Calculate the cost price of the camera.
SOLUTION
Cost price $= 84 \times \frac{100}{112} = \$75$

7. Matthew makes pieces of furniture and sends them to a shop where they are sold. When a piece is sold, the shopkeeper receives 15% of the selling price, and Matthew receives the rest.

(a) A table is sold for $200.
(i) Calculate the amount the shopkeeper receives.
$\frac{15}{100} \times 200 = \$30$
(ii) The cost of making this table was $131.80. Calculate the percentage profit that Matthew makes when this table is sold.

Profit $= \$200 - \$131.80 - \$30 = \$38.20$
Percentage profit $= \frac{38.20}{131.80} \times 100 = 28.98\%$

(b) Matthew made a bookcase.
The cost of making the bookcase was $647.50. After the bookcase is sold and the shopkeeper has received 15% of the selling price, Matthew makes a profit of $160. Calculate the selling price of the bookcase.

Selling price $= (647.50 + 160) \times \frac{100}{85} = \$950$

CHAPTER 2 : PERCENTAGES SOLUTIONS

CLASS ACTIVITY

Complete the table below.

Cost Price	Selling Price	Profit/Loss	Percentage profit/ Percentage Loss
$20	$24	Profit = $4	% profit = $\frac{profit}{cost\ price} \times 100$ $= \frac{4}{20} \times 100 = 20\%$
$200	$190	Loss = $10	% loss = $\frac{10}{200} \times 100 = 5\%$
$460	$598	Profit = $138	% profit = $\frac{138}{460} \times 100 = 30\%$
$50	$23	Loss = $27	% loss = $\frac{27}{50} \times 100 = 54\%$

Find the selling price in each of the following cases.

1. Cost price $60, profit 15%

    Selling price = $1.15 \times 60 = \$69$

2. Cost price $500, loss 8%

    Selling price = $0.92 \times 500 = \$460$

3. Cost price $1200, profit 2.5%

    Selling price = $1.025 \times 1200 = \$1230$

4. Cost price $5400, loss 3%

    Selling price = $0.97 \times 5400 = \$5238$

5. Cost price $2000, profit 5%

    Selling price = $1.05 \times 2000 = \$2100$

6. Cost price $72, loss 4%

    Selling price = $0.96 \times 72 = \$69.12$

Find the cost price in each of the following cases.

7. Selling price $550, profit 10%

    Cost price = $550 \times \frac{100}{110} = \$500$

8. Selling price $558, profit 24%

    Cost price = $558 \times \frac{100}{124} = \$450$

9. Selling price $680, loss 10%

    Cost price = $680 \times \frac{100}{90} = \$755.56$

10. Selling price $1200, profit 20%

    Cost price = $1200 \times \frac{100}{120} = \$1000$

MATHEMATICS APPLICATIONS UNIT 1

EXERCISE 2E

1. A man bought a picture for $325 and sold it at a profit of 12%. Calculate the selling price.

    $1.12 \times 325 = \$364$

2. A dealer made a profit of 20% by selling a car for $6300. Calculate the price he paid for the car.

    $6300 \times \frac{100}{120} = \$5250$

3. The owner of a toy shop made a profit of 40% on every toy which he sold. Find the selling price of a soccer ball which cost $11.

    $1.40 \times 11 = \$15.40$

4. Jordan usually sells refrigerators for $96 each. He discovers that one refrigerator has been slightly damaged, so he reduces the price by 15%.

    (a) Calculate the new selling price for the damaged refrigerator.

    $96 \times 0.85 = \$81.60$

    (b) When he sells the refrigerator at the usual price of $96, Jordan makes a profit of 20%. Calculate the cost price of the refrigerator.

    $96 \times \frac{100}{120} = \$80$

    (c) The shopkeeper made a profit of $10 when he sold a kite. Calculate the selling price of the kite.

    $10 \times \frac{100}{40} = \$25$

(b) The cost price of a Barbie doll which was sold for $28.

$28 \times \frac{100}{140} = \$20$

5. In a sale, a shop reduces all its prices by 20%.

    (a) Calculate the cost of an article which was originally priced at $26.

    $26 \times 0.8 = \$20.80$

    (b) The original price of an article which was sold in the sale for $44.

    $44 \times \frac{100}{80} = \$55$

6. A bookseller bought a gardening book for $7.50 and sold it for $10.50. Calculate the percentage profit made by the bookseller.

    $\frac{3}{7.50} \times 100 = 40\%$

    (b) During a sale the bookseller reduced the price of his books by 30%. Calculate the normal selling price of a book which was priced in the sale at $13.30.

    $13.30 \times \frac{100}{70} = \$19$

CHAPTER 2 : PERCENTAGES SOLUTIONS

2F DISCOUNT

Discount is a reduction from the usual cost of a product. For example, some customers get a discount for buying in bulk. Some businesses give discount to attract customers to buy more and thus increasing their sales.

EXAMPLES

1. A new television set cost $800. For the month of June a discount of 5% is allowed to all buyers. Calculate the cost of the television in June.

SOLUTION

$$800 \times 0.95 = \$760$$

2. Joe paid $108 for a set of books after a discount of 10%. Calculate the original price of the set of books.

SOLUTION

$$108 \times \frac{100}{90} = \$120$$

EXERCISE 2F

Find the selling price of each of the following after the discount.

1. Original price $200, 15% discount

$$200 \times 0.85 = \$170$$

2. Original price $300, 12% discount

$$300 \times 0.88 = \$264$$

3. Original price $600, 5% discount

$$600 \times 0.95 = \$570$$

4. Original price $2000, 2.5% discount

$$2000 \times 0.975 = \$1950$$

5. Original price $450, 24% discount

$$450 \times 0.76 = \$342$$

6. Original price $8200, 4% discount

$$8200 \times 0.96 = \$7872$$

Find the original price in each of the following.

7. A sum of $132 is paid after a discount of 12%.

$$132 \times \frac{100}{88} = \$150$$

8. A sum of $3551.20 is paid after a discount of 8%.

$$3551.20 \times \frac{100}{92} = \$3860$$

9. A sum of $23040 is paid after a discount of 4%.

$$23040 \times \frac{100}{96} = \$24000$$

10. A sum of $2375 is paid after a discount of 5%.

$$2375 \times \frac{100}{95} = \$2500$$

11. A shop is offering discounts on shirts costing $30 each. If someone buys 2 shirts, he will be offered a discount of 12% on the first shirt and another 10% discount on the reduced price for the second shirt. How much would one pay for two shirts at this shop?

$first\ shirt: 30 \times 0.88 = \$26.40$

$second\ shirt: 26.40 \times 0.9 = \$23.76$

$Total: 26.40 + 23.76 = \$50.16$

12. The price of a pair of Nike socks was decreased by 22% to $30. What was the original price of the socks?

$$30 \times \frac{100}{78} = \$38.46$$

13. (a) In 2013 the cost of posting a letter was 36 cents.

(i) A company posted 3000 letters and was given a discount of 4%. Calculate the total discount given. Give your answer in dollars.

$$3000 \times 0.36 \times 0.04 = \$43.20$$

(ii) In 2014, the cost of posting a letter was increased from 36 cents to 45 cents. Calculate the percentage increase.

$$\frac{9}{36} \times 100 = 25\%$$

(iii) After the price increase to 45 cents, the cost to the company of posting 3000 letters was $1302.75. Calculate the new percentage discount given.

$3000 \times 0.45 = \$1350$

$discount: 1350 - 1302.75 = \$47.25$

$percentage\ discount: \frac{47.25}{1350} \times 100 = 3.5\%$

(b) In 2014, it cost $5.40 to post a parcel. This was an increase of 12.5 % on the cost of posting the parcel in 2013. Calculate the increase in the cost of posting this type of parcel in 2014 compared to 2013.

$cost\ in\ 2013 = 5.40 \times \frac{100}{112.5} = \$4.80$

$Increase\ in\ cost = 5.40 - 4.80 = 60c$

CHAPTER 2 : PERCENTAGES SOLUTIONS

2G COMMISSION

Commission is money that a person earns based on how much he or she sells. It is usually expressed as a percentage of the total amount sold.

The following people normally work on commission as well as a retainer (fixed wage) : Real estate agents, car dealers, and pharmaceutical sales representatives. For example, a real estate agent earns a fraction of the selling price of a house that he or she helps a customer to purchase or sell. Similarly, a car dealer earns a portion of the selling price of an automobile that he or she is able to sell.

EXAMPLE 1

Patrick, an art dealer, earns 12% commission of the dollar value of the art pieces that he sells at the Louvre. Patrick sold $42000 worth of art pieces this month. How much commission does he receive?

SOLUTION

Commission = $0.12 \times 42000 = \$5040$

EXAMPLE 2

Alex is a car dealer and he earns 15% commission of his exclusive range of luxury vehicles sales. Last year his salary was $48000. What was the total dollar amount of his sales last year?

SOLUTION

Total sales = $48000 \times \dfrac{100}{15} = \$320000$.

EXAMPLE 3

George is employed by a real-estate agency and is paid commission as follows:

First $ 200 000 $\rightarrow$ 4 %

Amount exceeding $ 200 000 $\rightarrow$ 2.5%

George sold a property worth $ 280 000. Calculate the total commission paid to him.

SOLUTION

First $ 200 000 $\rightarrow 0.04 \times 200\ 000 = \$ 8000$

Amount exceeding $ 200 000 (280 000 − 200 000 = 80 000) $\rightarrow 2.5\% \times 80\ 000 = \$ 2000$

Total commission = $ 8000 + $ 2000 = $ 10 000

EXERCISE 2G

1. Chloe sells $2500 worth of toys, and she makes 20% commission. How much money did she make? $$0.20 \times 2500 = \$500$$	**2.** Dane's weekly pay is $480, and he gets 10% commission on his sale. If he sells $2800 worth of goods, how much money does he get in total? $$0.10 \times 2800 = \$280$$ $$total\ pay = \$280 + 480 = \$760$$
3. A salesman earns $3000 per month plus 2% commission on his sales. If he sells goods worth $24000 in one month, how much does he earn in that month? $$0.02 \times 24000 = \$480$$ $$he\ earns\ \$480 + 3000 = \$3480$$	**4.** Jenna is a salesperson at an electronics store. She earns 6.5 % commission on her total sales. What would be her commission if she sold a $2950 plasma television set? $$0.065 \times 2950 = \$191.75$$
5. Clara makes $800 per month plus 8% commission. What must commission sales reach for her total salary to be $2000? $$2000 - 800 = 1200$$ $$1200 \times \dfrac{100}{8} = \$15000$$	**6.** Alexandra makes $7 an hour plus 12% commission selling jewellery. How much does she earn on an 8 hour shift in which she sells $450 worth of jewellery? $$0.12 \times 450 = \$54$$ $$she\ earns\ \$54 + 8 \times 7 = \$110$$
7. John, a computer salesperson earns $35 000 as a retainer each year. This year John sold $240 000 worth of computers. If a commission of 2.5% is paid on the sales, find John's income for the year. $$0.025 \times 240000 = \$6000$$ $$John\ earns\ \$6000 + 35000 = \$41000$$	**8.** Peter receives a flat salary of $3000 a month plus a commission of 2% for the value of goods he sells. During a particular month he received $3360 in total. Calculate the total value of goods he sold that month. $$3360 - 3000 = 360$$ $$he\ sold\ \ 360 \times \dfrac{100}{2} = \$18000$$

CHAPTER 2 : PERCENTAGES SOLUTIONS

9. Jean Luc, an art dealer, earns 15% commission of the dollar value of the art pieces that he sells at the Australian Academy of Arts. Jean Luc earns $9600 this month. What is the total dollar value of the art that he sells?

$$9600 \times \frac{100}{15} = \$64000$$

10. Roger is an agent for movie stars. He earns 12% of his clients' salaries. If he made $75,000 last year, how much did his clients make in all?

$$75000 \times \frac{100}{12} = \$625000$$

11. Matilda works as a salesgirl in a jewellery store on Murray Street. She is paid a 9.5% commission on her sales. One very busy day she made the following 3 sales;
A ladies watch: $275.95
A diamond necklace: $599.99
A pair of cufflinks: $125
What was Matilda's commission on her total sales?

$$Total\ sales\ \ 275.95 + 599.99 + 125 = \$1000.94$$

$$Commission = 0.095 \times 1000.94 = \$95.09$$

12. George is employed by a real-estate agency and is paid commission as follows :

First $ 200 000 → 2.5 %
Next $ 100 000 → 2 %
Amount exceeding $ 300 000 → 1.5%

George sold a property worth $ 360 000. Calculate the amount of commission he earned.

$0.025 \times 200000 = \$5000$
$0.02 \times 100000 = \$2000$
$0.015 \times 60000 = \$900$
$Commission = 5000 + 2000 + 900$
$= \$7900$

13. A salesperson receives step commission on sales calculated as follows:

* 8% on first $1000
* 10% next $2000
* 15% on sales above $3000

Calculate the salesperson's earning if in one week his sales was $8500.

$0.08 \times 1000 = \$80$
$0.1 \times 2000 = \$200$
$0.15 \times 5500 = \$825$
$Commission = 80 + 200 + 825 = \$1105$

2H CALCULATING TAX PAID AND TAXABLE INCOME

EXAMPLE

The current income tax rates for individuals in Australia are shown in the table below.

Taxable income	Tax on this income
0 - $18 200	Nil
$18 201 - $37 000	19c for each $1 over $ 18 200
$37 001 - $80 000	$3 572 plus 32.5c for each $1 over $37 000
$80 001 - $180 000	$17 547 plus 37c for each $1 over $80 000
$180 001 and over	$54 547 plus 45c for each $1 over $180000

Use the table to determine the income tax paid by an individual who has a taxable income of

(a) $15500

$$tax\ paid = \$0$$

(b) $27500

$$Tax\ paid = 0.19(27500 - 18200) = \$1767$$

(c) $85400.

$$Tax\ paid = 17547 + 0.37(85400 - 80000) = \$19545$$

(d) How much does the individual in part (c) take home monthly?

$$\frac{85400 - 19545}{12} = \$5487.92$$

(e) Julian paid $15064 in tax. Determine Julian's taxable income.

Clearly Julian must in the $37 001 - $80000 tax brackets.

Solve $(3572 + 0.325(x - 37000) = 15064)$ on CAS

$$x = \$72360$$

Therefore Julian's taxable income is $72360.

CHAPTER 2 : PERCENTAGES SOLUTIONS

EXERCISE 2H

Income tax rates for 2014–15 for both Australian citizens and foreign residents working in Australia are given in the tables below. The rates apply from 1 July 2014.

Taxable income	Tax on this income
0 – $18 200	Nil
$18 201 – $37 000	19c for each $1 over $18 200
$37 001 – $80 000	$3 572 plus 32.5c for each $1 over $37 000
$80 001 – $180 000	$17 547 plus 37c for each $1 over $80 000
$180 001 and over	$54 547 plus 45c for each $1 over $180000

1. Use the table to determine the income tax paid by Alex who has a taxable income of $54500. $Tax\ paid$ $= 3572 + 0.325(54500 - 37000)$ $= \$9259.50$	2. James has a taxable income of $105 000. Determine the income tax he paid. $17547 + 0.37(105000 - 80000)$ $= \$26797$
3. Use the table to determine the income tax paid by Charles who has a taxable income of $32000. $Tax\ paid = 0.19(32000 - 18200)$ $= \$2622$	4. Dave has a taxable income of $190000. Use the table to determine the income tax he pays. $54547 + 0.45(190000 - 180000)$ $= \$59047$
5. Calculate the taxable income of Eric who paid $23615 in tax in the 2014–15 financial year. Solve $(17547 + 0.37(x - 80000)$ $= \$23615, x)$ $x = 96400$ ∴ *Eric's taxable income is $96400*	6. Calculate the taxable income of Fredric who paid $61927 in tax in the 2014–15 financial year. $54547 + 0.45(x - 180000) = \$61927$ $x = 196400$ ∴ *Fredric's taxable income is $196400*

7. From 1 July 2013, these fees were recommended by the Roswell Real Estate Company.

Real Estate Fees/Commission in Roswell

Selling price of each property	Real estate fees
Does not exceed $10 000	10.2% with a minimum of $100
From $10 001 – $50 000	$1050 plus 5.8% of excess over $10 000
From $50 001 – $100 000	$3400 plus 4.1% of excess over $50 000
From $100 001 upward	$5460 plus 3.9% of excess over $100 000

Alexa owns several properties in Roswell.

(a) If Alexa sells a house for $280 000, how much does she have to pay in real estate fees?

$$5460 + 0.039 \times (280\,000 - 100\,000) = \$12480$$

(b) In a particular week Alexa sells two properties, a house for $198 000 and a unit for $99 900. Calculate the total fees she needs to pay.

$$5460 + 0.039 \times (198\,000 - 100\,000) = \$9282$$
$$3400 + 0.041 \times (99900 - 50\,000) = \$5445.90$$
$$Total\ fees = 9282 + 5445.90 = \$14727.90$$

(c) Alexa wants to reduce the number of properties she owns. She is trying to decide whether to sell a house for $280 000 or sell a group of four apartments, for $70 000 each. Which option will result in her paying the smaller amount of real estate fees and how much less will she pay in fees with this option?

$Fees\ paid\ on\ \$70\,000\ apartment = 3400 + 0.041(70000 - 50000) = \$4220$
$Fees\ paid\ on\ four\ apartments = 4 \times 4220 = \$16880$
She'll pay $16880 - 12480 = \$4400\ less\ by\ selling\ a\ \$280\,000\ house.$

CHAPTER 3

SIMPLE INTEREST

3A CALCULATING SIMPLE INTEREST

To find the simple interest (I), use the formula I = P R T, where P is the principal value or present value, R is the rate of interest and T is time in years.

EXAMPLE 1

Find the simple interest to be earned on each of these investments.

(a) $4000 for 3 years at 9% p.a.

$$4000 \times \frac{9}{100} \times 3 = \$ 1080$$

(b) $2000 for 9 months at 6.5% p.a.

$2000 \times \frac{6.5}{100} \times 0.75 = \$ 97.50$ [9 months is ¾ of a year or simply divide 9 by 12]

EXAMPLE 2

A family borrows $16000 to buy a car and are charged 15% p.a. simple interest. If the loan is for 5 years find:

(a) the amount of interest to be paid

PRT = 16000 × 0.15 × 5 = $ 12000

(b) the total amount to be repaid

$16000 + $ 12000 = $ 28000

(c) the amount of each repayment if the loan is to be repaid in 60 equal instalments.

28000 ÷ 60 = $ 466.67

EXAMPLE 3

Peter intends to buy a new car costing $24000 for his 21st birthday. He saved 30% of the amount needed in the bank. He decided to borrow the remainder from MeBank and agree to make quarterly repayments, with simple interest to be paid at the rate of 12% per annum. Including the principal, Peter will pay the debt in full in 5 years. Calculate Peter's quarterly repayments.

SOLUTION

Amount borrowed = 70% of $ 24000 = $ 16800

Interest payable = 16800 × 0.12 × 5 = $10080

Total amount to be repaid = 16800 + 10080 = 26880

5 years = 5 × 4 = 20 quarters ∴ Quarterly payments = 26880 ÷ 20 = $ 1344

MATHEMATICS APPLICATIONS UNIT 1

EXERCISE 3A

Find the simple interest to be earned on each of these investments.

1. $500 for 2 years at 16% p.a. **$160**	2. $2 500 for 4 years at 1.2% p.a. **$120**
3. $16 450 for 3 years at 8% p.a. **$3948**	4. $350 for 210 days at 42% p.a. $350 \times 0.42 \times \frac{210}{365} = \mathbf{\$84.58}$
5. How much interest is payable if I borrow $6 000 at 7.4% for 21 months? $6000 \times 0.074 \times \frac{21}{12} = \mathbf{\$777}$	7. A family borrows $850 to buy a television and are charged 18% p.a. simple interest. If the loan is for 2 years find:
6. $15 000 is invested at 6% p.a. simple interest for 10 years.	**a** the amount of interest to be paid **$306**
a How much interest will be earned each year? **$900**	**b** the total amount to be repaid $850 + 306 = \mathbf{\$1156}$
b How much interest will be earned over the 10-year period? **$9000**	**c** the amount of each repayment if the loan is to be repaid in 24 equal instalments. $\frac{1156}{24} = \mathbf{\$48.17}$

CHAPTER 3 : SIMPLE INTEREST SOLUTIONS

8.

Account	Simple interest per year
Alpha saver	4.2%
Beta saver	4.4%

On 31 December 2011, Kelly and Johan each had $6000 in an account.

Kelly's money is in a Alpha Saver Account. Johan's money is in a Beta Saver Account.

(i) How much money did Kelly have in her account on 31 December 2012 after the interest had been added?

$$I = 6000 \times 0.042 \times 1 = \$252$$

Kelly has $6000 + 252 = \$6252$

(ii) On 31 December 2012, Kelly transferred this money to the Beta Saver Account. How much money did she have in this account on 31 December 2013 after the interest had been added?

$$I = 6252 \times 0.044 \times 1 = \$275.09$$

Kelly has $6252 + 275.09 = \$6527.09$

(iii) Johan kept her money for the two years in the Beta Saver Account, which earned simple interest of 4.4% per year.
After all interest had been added, who had more money in their account on 31 December 2013 and by how much?

$$I = 6000 \times 0.044 \times 2 = \$528$$

Johan has $6000 + 528 = \$6528$

Johan has 91c more than Kelly

3B SIMPLE INTEREST AND TECHNOLOGY (CAS)

Simple interest can also be calculated by using the class pad calculator.

• Menu → Financial → Simple interest

EXAMPLE 1

Use the class pad to find the interest obtained when $6000 is invested for 3 years at 8% per annum.

SOLUTION

Days	$365 \times 3 = 1095$
I%	8
PV	6000
SI	Solve and 1440 appears (**ignore the minus sign**)
SFV	If you solve this gives $6000 + 1440 = 7440$ which is the interest added to the principal value

EXERCISE 3B

Using your calculator, find the simple interest to be earned on each of these investments.

1. $2500 for 4 years at 12% p.a. **$1200**	2. $4 500 for 2 years at 6.5% p.a. **$585**
3. $16 300 for 120 days at 8% p.a. **$428.71**	4. $45000 for 5 years at 8.5% p.a. **$19125**
5. Calculate the simple interest when $5400 is invested for 6.5 years at 8.8% per annum. **$3088.80**	6. Calculate the simple interest when $6500 is invested for 18 months at 7.6% per annum. $6500 \times 0.076 \times 1.5 = \$741$

3C FINDING (P), RATE OF INTEREST (R) OR TIME (T)

If three of the four variables in the formula I = P R T are known, we can calculate the fourth one by simply rearranging the formula or using CAS as show in the examples below.

$$P = \frac{100\,I}{R \times T}, \quad R = \frac{100\,I}{P \times T} \quad \text{and} \quad T = \frac{100\,I}{P \times R}$$

EXAMPLE

Joseph invested $3000 for 4 years and his savings accumulated to $3600. Calculate the rate of interest.

SOLUTION

Alternative 1

Using $R = \frac{100\,I}{P \times T} = \frac{100 \times 600}{3000 \times 4} = 5$

Rate of interest = 5%

Alternative 2 (CAS)

Main→ Action →Advanced → Solve
Using I = P R T
Solve (600 = 3000 × x × 4)
x = 0.05
rate of interest = 5%

EXERCISE 3C

Find the principal invested in each of *these* simple interest investments.

1. Interest of $400 at 8% for 5 years.

   $P = \frac{100\,I}{R \times T} = \frac{100 \times 400}{8 \times 5}$

   $= \$1000$

2. Interest of $6200 at 3.2% for 2 years

   $96875

3. Interest of $5 000 at 7% p.a. for 2 years

   $35714.29

4. Interest of $1 400 at 9.2% for 3 years

   $5072.46

5. An investor has $50 000 invested in an account that pays 4.6% p.a. simple interest. If she wants to earn at least $20 000 in interest, for how many years will the money need to be invested?

   $20000 = 50000 \times 0.046 \times x$
   (CAS : Main, Action, Advance, Solve)

   $x = 8.7\ years$

6. Darren has received $90 000 in interest payments on an investment of $500 000 that he made 4 years ago which paid simple interest. What rate of interest has been paid?

   $R = \frac{100 \times 90000}{500000 \times 4} = 4.5\%$

7. A sum of $350 is invested at simple interest and amounts to $476 after 3 years. Calculate
   (a) The total interest earned,

   $476 - 350 = \$126$

   (b) The rate percent per annum.

   $R = \frac{100\,I}{P \times T}$

   $= \frac{100 \times 126}{350 \times 3} = 12\%$

8. Calculate the simple interest when $2800 is invested for 9 months at 8% per annum.

   $9 \div 12 = 0.75$

   $SI = 2800 \times 0.08 \times 0.75 = \$168$

9. Lily, Margaret and Nancy were each left $8000 in their aunt's will. Margaret invested her money in a bank at 9.5% simple interest for 5 years. Calculate the total amount of money she had in the bank after 5 years.

   I = P R T

   $8000 \times 0.095 \times 5 = \$3800$

   $Total\ amount = 8000 + 3800 = \$11800$

10. Find the simple interest obtained when $12500 is invested at 7.2% for a period of 27 months.

    $27 \div 12 = 2.25$

    $SI = 12500 \times 0.072 \times 2.25$

    $= \$2025$

11. A bank charges $28 simple interest on a sum of money which is borrowed for four months.
    Given that the rate of interest is 15% per annum, calculate the sum of money.

    $P = \frac{100\,I}{R \times T} = \frac{100 \times 28}{15 \times \frac{1}{3}} = \$560$

12. How long does it take $12000 to yield a simple interest of $2640 at 5.5% per annum?

    $T = \frac{100\,I}{P \times R} = \frac{100 \times 2640}{12000 \times 5.5}$

    $= 4\ YEARS$

13. What sum of money will yield a simple interest of $5400 at 9% per annum in 3 years?

    $P = \frac{100\,I}{R \times T} = \frac{100 \times 5400}{9 \times 3}$

    $= \$20000$

14. Joseph has a deposit of $16 000 in an investment account which pays a simple interest rate of 5% p.a. How long would it take him to earn $5 200 in interest?

    $Solve\ (5200 = 16000 \times 0.05 \times x)$

    $x = 6.5\ years$

CHAPTER 4

COMPOUND INTEREST

4A CALCULATING COMPOUND INTEREST

Compound interest is different from Simple Interest as it involves adding interest at regular intervals to the amount invested or borrowed.

The formula that can be used to find the compound interest is given by

$$A = P\left(1 + \frac{R}{n \times 100}\right)^{n \times T}$$

- A is the principal plus interest
- P means Principal value,
- R is the rate of interest,
- T is time in years and
- n is the number of times interest is compounded per year. For example, if interest is compounded quarterly n = 4.

EXAMPLES

1. Emily invests $2450 in a bank for 4 years at 4.8% per annum, where the interest is compounded. Find the **value of her investment**, if the interest is paid:

(a) annually

$$A = 2450\left(1 + \frac{4.8}{1 \times 100}\right)^{1 \times 4}$$
$$= \$2955.37$$

(b) monthly

$$A = 2450\left(1 + \frac{4.8}{12 \times 100}\right)^{12 \times 4}$$
$$= \$2967.46$$

(c) every six months

$$A = 2450\left(1 + \frac{4.8}{2 \times 100}\right)^{2 \times 4}$$
$$= 2961.87$$

2. Robredo invests $10000 in a bank for 6 years at 8% per annum, where the interest is compounded. Find the **interest earned** on his investment, if the interest is paid:

(a) quarterly

$$A = 10000\left(1 + \frac{8}{4 \times 100}\right)^{4 \times 6}$$
$$= \$16084.37$$
$$\text{Interest} = \$16084.37 - 10000$$
$$= \$6084.37$$

(b) daily

$$A = 10000\left(1 + \frac{8}{365 \times 100}\right)^{365 \times 6}$$
$$= \$16159.89$$
$$\text{Interest} = \$16159.89 - 10000$$
$$= \$6159.89$$

EXERCISE 4A

1. Alex invests $9600 in a bank for 5 years at 9% per annum, where the interest is compounded. Find the value of his investment, if the interest is paid:

(a) Annually

$$A = 9600\left(1 + \frac{9}{1 \times 100}\right)^{1 \times 5}$$
$$= \$14770.79$$

(b) each 6 months

$$A = 9600\left(1 + \frac{9}{2 \times 100}\right)^{2 \times 5}$$
$$= \$14908.51$$

(c) each month

$$A = 9600\left(1 + \frac{9}{12 \times 100}\right)^{12 \times 5}$$
$$= \$15030.54$$

(d) quarterly

$$A = 9600\left(1 + \frac{9}{4 \times 100}\right)^{4 \times 5}$$
$$= \$14980.89$$

2. Susan invests $12000 in a bank for 3 years at 10% per annum, where the interest is compounded. Find the **interest earned** on her investment, if the interest is paid:

(a) Annually

$$A = 12000\left(1 + \frac{10}{1 \times 100}\right)^{1 \times 3}$$
$$= \$15972$$
$$\text{Interest} = \$15972 - 12000$$
$$= \$3972$$

(b) each 6 months

$$A = 12000\left(1 + \frac{10}{2 \times 100}\right)^{2 \times 3}$$
$$= \$16081.15$$
$$\text{Interest} = \$16081.15 - 12000$$
$$= \$4081.15$$

(c) each month

$$A = 12000\left(1 + \frac{10}{12 \times 100}\right)^{12 \times 3}$$
$$= \$16178.18$$
$$\text{Interest} = \$16178.18 - 12000$$
$$= \$4178.18$$

(d) quarterly

$$A = 12000\left(1 + \frac{10}{4 \times 100}\right)^{4 \times 3}$$
$$= \$16138.67$$
$$\text{Interest} = \$16138.67 - 12000$$
$$= \$4138.67$$

CHAPTER 4: COMPOUND INTEREST SOLUTIONS

4B COMPOUND INTEREST AND TECHNOLOGY

Compound interest can also be calculated by using the class pad. Use the following steps:

- Menu
- Financial
- Compound interest

EXAMPLE 1	EXAMPLE 2
Peter invests $6000 in a bank for 5 years at 9% per annum, where the interest is compounded. Find the value of his investment and the interest earned, if the interest is paid monthly. **SOLUTION** To obtain the meaning of each of the letters, click help	Ashley invests $5600 in a bank for 3 years at 8.4% per annum, where the interest is compounded. Find the value of his investment and the interest earned, if the interest is paid quarterly. **SOLUTION**
N $12 \times 5 = 60$	N $4 \times 3 = 12$
I% 9	I% 8.4
PV 6000	PV 5600
PMT 0	PMT 0
FV SOLVE and we get 9394.09 (The answer would appear negative, ignore it)	FV SOLVE and we get 7186.16 (The answer would appear negative, ignore it)
P/Y 12	P/Y 4
C/Y 12	C/Y 4
The value of his investment is $9394.09 and the interest earned = $9394.09 − 6000 = $3394.09	The value of her investment is $7186.16 and the interest earned = $7186.16 − 5600 = $1586.16

EXERCISE 4B

Use your class pad calculator (**financial**) to answer the following questions

1. John invests $8500 in a bank for 6 years at 12% per annum, where the interest is compounded. Find the value of his investment, if the interest is paid:

   (a) annually $16777.49

   (b) quarterly $17278.75

2. Peter borrows $12500 for 4 years at 9.6% per annum, where the interest is compounded. Find the interest paid if the interest is paid:

   (a) annually $5536.50

   (b) quarterly $5768.77

   (c) monthly   —

3. Alex invests $64000 in a bank for 8 years at 6.5% per annum, where the interest is compounded. Find the value of his investment, if the interest is paid:

   (a) quarterly $107200.76

   (b) monthly $107498.81

4. John wishes to invest $60 000 for 5 years. Three options are being considered.

   Option A: Simple interest at 8.6 % per year

   $$60000 \times 0.086 \times 5 = \$25800$$

   Option B: Compound interest at 7% per annum compounded yearly

   $$60000 \left(1 + \frac{7}{1 \times 100}\right)^{1 \times 5} = \$84153.10$$

   Interest = $84153.10 − 60000 = $24153.10

   Option C: Compound interest at 7.5% per annum compounded quarterly

   $$60000 \left(1 + \frac{7.5}{4 \times 100}\right)^{4 \times 5} = \$86996.88$$

   Interest = $86996.88 − 60000 = $26996.88

   Which investment should John choose to maximise the interest earned?   **Option C**

CHAPTER 4 : COMPOUND INTEREST SOLUTIONS

5. Sarah invests $1 250 in a bank for 3 years at 4.8% per annum, where the interest is compounded. Find the value of her investment, if the interest is paid quarterly.

$$A = 1250\left(1 + \frac{4.8}{4 \times 100}\right)^{4 \times 3}$$
$$= \$1442.37$$

6. The West Bank offers an account in which compound interest is calculated every quarter. The interest rate is 5.2 % per annum. The South Bank pays 5.4% compound interest and is calculated yearly. Philip has inherited $7000 from his grandmother's will and plans to invest the money for a period of 5 years. Which bank will offer him more interest?

    WEST BANK
    $A = 7000\left(1 + \frac{5.2}{4 \times 100}\right)^{4 \times 5}$
    $= \$9063.31$
    $Interest = \$2063.31$

    SOUTH BANK
    $A = 7000\left(1 + \frac{5.4}{1 \times 100}\right)^{1 \times 5}$
    $= \$9105.44$
    $Interest = \$2105.44$

    South Bank offers more interest by $42.13

7. Casey has $115 000 and wishes to invest it for 4 years. She checks out two banks and is offered the following terms:

    Bank A: 5.5% p.a. compounded yearly
    Bank B: 5.3 % p.a. compounded monthly.
    Determine which bank offers the best deal.

    BANK A
    $A = 115000\left(1 + \frac{5.5}{1 \times 100}\right)^{1 \times 4}$
    $= \$142464.83$

    BANK B
    $A = 115000\left(1 + \frac{5.3}{12 \times 100}\right)^{12 \times 4}$
    $= \$142090.66$

    Bank A offers the best deal by $374.17

MATHEMATICS APPLICATIONS UNIT 1

4C   HOW MUCH DO I REPAY?

We can also use the inbuilt financial programmes to determine the amount to be repaid by an individual when the latter borrows from any financial institution. The steps are same as before.

- Menu
- Financial
- Compound interest

EXAMPLE

Cyrus borrowed $20000 with Delta Bank with compound interest of 14% per annum, the amount to be paid off in 5 years. Calculate the constant amount to be repaid (PMT) each month for Cyrus to see his loan to be written off.

SOLUTION

N	$12 \times 5 = 60$
I%	14
PV	20000
PMT	Solve to obtain - $465.37
FV	0
P/Y	12
C/Y	12

Cyrus has to make a monthly repayment of $465.37.

EXERCISE 4C

1.	Amanda borrowed $15000 with Alpha Bank with compound interest of 16% per annum, the amount to be repaid off in 4 years. Calculate the constant amount to be repaid each month for Amanda to see her loan written off. $425.10	2.	Calculate to the nearest cent the constant monthly repayment if $240 000 is loaned for 20 years with compound interest of 6.1% per year. $1733.31
3.	Calculate to the nearest cent the constant weekly repayment if $60000 is loaned for 10 years with compound interest of 9.5% per year. $178.84	4.	Calculate to the nearest cent the constant yearly repayment if $8000 is loaned for 6 years with compound interest of 18% per year. $2287.28

4D INFLATION

Inflation can be defined as the rate at which the general level of prices for goods and services is rising, and, subsequently, purchasing power is falling. Central banks in most countries attempt to minimise severe inflation, along with severe deflation, in an attempt to keep the excessive growth of prices to a minimum.

CLASS ACTIVITY : Copy and complete the table below.

Rate of inflation	Workings	Used in answering questions
5%	100 + 5 = 105%	1.05
7%	100 + 7 = 107%	1.07
13%	100 + 13 = 113%	1.13
20%	100 + 20 = 120%	1.20
4.8%	100 + 4.8 = 104.8%	1.048
8.8%	100 + 8.8 = 108.8%	1.088
5.25%	100 + 5.25 = 105.25%	1.0525

EXAMPLE 1

A gold necklace is bought for $3000 and gains 6% of its value each year. What is the value after:

(i) 3 years   (ii) 10 years   (iii) n years

SOLUTION

(i) $3000(1.06)^3 = \$3573.05$
(ii) $3000(1.06)^{10} = \$5372.54$
(iii) $3000(1.06)^n$

EXAMPLE 2

An item is bought for $5000 and is valued at $6800 after 5 years. Find the constant rate of inflation that would allow this rise in value.

SOLUTION

$5000 x^5 = 6800$ (Main, Action, Advanced, Solve)

$x = 1.063$

Inflation rate = 6.3%

EXERCISE 4D

1. An item is bought for $4000 and gains 5% of its value each year.
What is the value after:

   (i) 4 years

   $4000(1.05)^4 = \$4862.03$

   (ii) 10 years

   $4000(1.05)^{10} = \$6515.58$

   (iii) n years

   $4000(1.05)^n$

2. A house is bought for $250 000 and gains 4.5% of its value each year. What is the value after:

   (i) 5 years

   $250000(1.045)^5 = \$311545.48$

   (ii) 15 years

   $250000(1.045)^{15} = \$483820.61$

   (iii) n years

   $250000(1.045)^n$

3. A piece of jewellery is bought for $1200 and gains 6.25% of its value each year. What is the value after 3 years?

   $1200(1.0625)^3 = \$1439.36$

4. An item is bought for $12000 and is valued at $14000 after 3 years. Find the constant rate of inflation that would allow this rise in value.

   $Solve\ (12000 \times x^3 = 14000, x)$

   (Action, advanced, Solve)

   $x = 1.053$

   Answer : 5.3%

5. An item is bought for $10000 and is valued at $12500 after 5 years. Find the constant rate of inflation.

   $Solve\ (10000 \times x^5 = 12500, x)$

   $x = 1.046$

   Answer : 4.6 %

6. An item is bought for $400 and doubles its value after 6 years. Find the constant rate of inflation.

   $Solve\ (400 \times x^6 = 800, x)$

   $x = 1.12246$

   Answer : 12.25 %

CHAPTER 4: COMPOUND INTEREST SOLUTIONS

4E DEPRECIATION

Depreciation is the decline in the value of an item due to wear and tear. For example, the value of a car normally depreciates overtime.

Copy and complete the table below.

Rate of depreciation	Workings	Used in answering questions
5%	100 − 5 = 95%	0.95
2%	100 − 2 = 98%	0.98
12%	100 − 12 = 88%	0.88
25%	100 − 25 = 75%	0.75
4.5%	100 − 4.5 = 95.5%	0.955
8.5%	100 − 8.5 = 91.5%	0.915
5.65%	100 − 5.65 = 94.35 %	0.9435%

EXAMPLE 1

A car is bought for $24 000 and loses 9% of its value each year. What is the value after:

(i) 3 years   (ii) 10 years   (iii) n years

SOLUTION

(i) $24000 (0.91)^3 = \$ 18085.70$
(ii) $24000 (0.91)^{10} = \$9345.99$
(iii) $24000 (0.91)^n$

EXAMPLE 2

An item is bought for $5000 and is valued at $3500 after 5 years. Find the constant rate of depreciation that would allow this drop in value.

SOLUTION

$5000 x^5 = 3500$ (Action, Advanced, Solve on your CAS calculator)

$x = 0.93$

Depreciation rate = 7%

EXERCISE 4E

1. A car is bought for $32000 and loses 8% of its value each year. What is the value after?

(i) 2 years

$$32000(0.92)^2 = \$27084.80$$

(ii) 9 years

$$32000(0.92)^9 = \$15109.16$$

(iii) n years

$$32000(0.92)^n$$

2. A machine is bought for $8000 and loses 13% of its value each year. What is the value after :

(i) 3 years

$$8000(0.87)^3 = \$5268.02$$

(ii) 5 years

$$8000(0.87)^5 = \$3987.37$$

(iii) n years

$$8000(0.87)^n$$

3. A PSP is bought for $1200 and loses 3.5% of its value each year. What is the value after 3 years?

$$1200(0.965)^3 = \$1078.36$$

4. An item is bought for $12000 and is valued at $8500 after 4 years. Find the constant rate of depreciation that would allow this drop in value.

$Solve(12000 \times x^4 = 8500, x)$

$x = 0.9174$

Answer : 8.26 %

5. An item is bought for $10000 and is valued at $7000 after 5 years. Find the constant rate of depreciation that would allow this drop in value.

$Solve(10000 \times x^5 = 7000, x)$

$x = 0.9311$

Answer : 6.89 %

6. An item halves its value after 3 years. Find the constant rate of depreciation that would allow this drop in value.

$Solve(200 \times x^3 = 100, x)$

$x = 0.7937$

Answer : 20.63 %

CHAPTER 5

FINANCIAL MATHEMATICS

5A CALCULATING WAGES FROM AN ANNUAL SALARY

In this section, the reader's objective is to work out the weekly, fortnightly or monthly wage given the annual salary. What is annual salary? Annual salary is a fixed amount of money or compensation paid to an employee by an employer in return for any kind of work performed over a period of one year. In simple words, annual salary is the amount a person is paid during the period of a year (12 months, 52 weeks, 26 fortnights).

EXAMPLES

1. Peter earns $48 000 annually. Calculate his monthly wage.
   SOLUTION
   Monthly wage = $\frac{48000}{12}$ = **$4000**.

2. John gets an annual salary of $62 500. What is his weekly wage?
   SOLUTION
   Weekly wage = $\frac{62500}{52}$ = **$1201.92**

EXERCISE 5A

1. The table below shows the income earned by 4 friends working for Beta Chemicals. Complete the table.

	Annual income ($)	Monthly income ($)	Weekly income ($)	Fortnightly income ($)
Alex	42000	3500	807.69	1615.38
Bob	37500	3125	721.15	1442.31
Carl	38740	3228.33	745	1490
Emma	25168	2097.33	484	968

2. Calculate the weekly income for each professional given their annual salary.

Head Teacher Salary $96 000	Football player Salary $248 000	Scientist Salary $86 500
$1846.15	$4769.23	$1663.46

3. Manuela earns a salary of $36460 per annum. If she works 42 hours in a week, how much does she get paid for that week?

   $36460 \div 52 = \mathbf{\$701.15}$

4. Maria's job pays her $1650 a month. What is her gross weekly wage?

   $1650 \div 4 = \mathbf{\$412.50}$

5. John grosses $29400 a year. What is his gross weekly wage?

   $29400 \div 52 = \mathbf{\$565.38}$

6. Sarah is a teacher and earns $48000 a year. Calculate her fortnightly gross pay.

   $48000 \div 26 = \mathbf{\$1846.15}$

7. Kevin works for ABC radio as an engineering tech and earns $45600 a year. Calculate his weekly gross pay.

   $45600 \div 52 = \mathbf{\$876.92}$

8. Jane works a 36 hour week and earns an annual salary of $48600. Her friend Anna works part time and earns $23.80 per hour.

   a How much does Jane earn each week?

   $48600 \div 52 = \mathbf{\$934.62}$

   b Who has the higher hourly rate of pay?

   Jane hourly rate = $934.62 \div 36 = \$25.96$
   ∴ **Jane has the higher hourly rate of pay**

   c If Anna works on average 24 hours per week, what is her yearly income?

   Anna's yearly income = $23.80 \times 24 \times 52 = \mathbf{\$29702.40}$

CHAPTER 5 : FINANCIAL MATHEMATICS SOLUTIONS

5B   CALCULATING WAGES FROM AN HOURLY RATE

Wage is basically how much a person is paid to do a job. It can be measured weekly, monthly or annually.

Hourly rate, on the other hand, is simply how much a person gets paid for an hour of work.

EXAMPLE 1
Luke works at Mc Donald and his hourly rate of pay is $11.50. Last week he worked for 12 hours. Calculate his basic pay.

SOLUTION

Wage $= 11.50 \times 12 = \$138$

EXAMPLE 2
Olivia starts work at 7.30am and finishes at 4 pm. She had a lunch break between 12 pm and 1 pm. Calculate her basic pay if she earns $10.75 per hour of work.

SOLUTION

7.30 am → 8.00 am → 12 pm → 1pm → 4 pm

30 mins   4 hours   3 hours

Total hours worked = 7.5 hours

Wage $= 7.5 \times 10.75 = \$80.63$

EXERCISE 5B

1. Peter works at Hungry Jacks and his hourly rate of pay is $10.80. Last week he worked for 14 hours. Calculate his basic pay. $10.80 \times 14 = \$151.20$	2. Becky starts work at 8.30 am and finishes at 3 pm. She had a lunch break between 12 pm and 12.30 pm. Calculate her basic pay if she earns $9.80 per hour of work. $6 \times 9.80 = \$58.80$
3. Shania drives a truck for $18.25 an hour and works 38 hours a week. What is her annual gross pay? $18.25 \times 38 \times 52 = \$36062$	4. Penny works as a waitress at the Cheese Cake Factory and makes $10.40/hour plus tips. Last week she worked 32 hours and made $463 in tips. What was her gross pay for the week? $10.40 \times 32 + 463 = \$795.80$
5. Richard is a word processor operator. He makes $14.50 an hour and works 32 hours a week. What would be his fortnightly gross pay? $14.50 \times 32 \times 2 = \$928$	6. Anastasia works a basic week of 40 hours and her basic rate is $9.80 per hour. Calculate her basic wage for the week. $9.80 \times 40 = \$392$
7. Jeremy works as a waiter in a Japanese restaurant. In addition to his regular pay of $11.20/hour, Jeremy keeps 85% of all the tips he receives. Calculate his gross weekly pay for a week in which he works 36 hours and receives $260 in tips. $11.20 \times 36 = \$403.20$ Tips kept $0.85 \times 260 = 221$ Weekly pay $= 403.20 + 221 = \$624.20$	

8. Ryan Biggs work full time four days a week at Target and the table below shows his Week 13 work schedule. He earns $12.40 per hour.

Name : Ryan Biggs			Employee No. 1831		Week 13
	In	Out	In	Out	Total hours worked
Mon	0900	1200	1300	1600	6
Tue	0800	1200	1400	1700	7
Wed	1030	1230	1300	1530	4.5
Fri	0800	1200	1400	1930	9.5
				Total	27

(a) Complete the table, stating the total hours worked each day.

(b) Hence, calculate his wage for the whole week.

$12.40 \times 27 = \$334.80$

5C OVERTIME AND OTHER ALLOWANCES

Overtime refers to doing extra work over and above your normal hours. Overtime allows people to earn a better hourly rate and thus more income. In this part of the chapter, emphasis will be laid on the two most common rates of overtime: double time ($\times 2$) and time and a half ($\times 1.5$).

EXAMPLES

1. Paul the junior plumber works for $22 per hour. His overtime rate is "double time". What does he get paid for 3 hours overtime?

SOLUTION

3 hours overtime = $3 \times 22 \times 2 = \$132$.

2. Jacob the painter works a basic 40 hours a week. He does 2 hours overtime at 'time and a half' on Saturday and 4 hours 'double time' on Sunday. His hourly rate is $18 per hour. Work out his total pay for the week.

SOLUTION

2 hours' time and a half
$= 2 \times 18 \times 1.5 = \$54$
4 hours double
$= 4 \times 18 \times 2 = \$144$
40 hours basic time
$= 40 \times 18 = \$720$
Total pay for the week
$= 54 + 144 + 720 = \$918$

3. Sam is employed at a hairdressing salon. He receives $16.50 per hour for a standard 35-hour Monday to Friday week. Sam also receives $45 per week as travel allowance and $200 per year as laundry allowance. Calculate Sam earnings for a standard week, assuming he receives his laundry allowance on a weekly basis.

SOLUTION

35 standard week = $35 \times 16.50 = \$577.50$
Travel allowance = $45
Laundry allowance = $200 \div 52 = 3.85$
Total earnings per week = $577.50 + 45 + 3.85 = \$626.35$

4. Courtney is paid time and a half for each hour she works over 32 hours in a week. Last week she worked 40 hours for a total of $726. What is her normal hourly rate?

SOLUTION

She worked 8 hours overtime
This is equivalent to $8 \times 1.5 = 12$ *hours* of normal hourly rate work
She worked $32 + 12 = 44$ *hours* at the normal hourly rate
Normal hourly rate = $\frac{726}{44} = \$16.50$

5. A secretary works a 35-hour week for which she is paid $444.50. In a particular week she works 4 hours overtime on Saturdays which is paid for at time-and-a-half, and 2.5 hours overtime on Sunday which is paid for at double-time, calculate her gross wage for that week.

SOLUTION

Hourly rate = $\frac{444.50}{35} = \$12.70$
Saturday's wage = $4 \times 1.5 \times 12.70 = \$76.20$
Sunday's wage = $2.5 \times 2 \times 12.70 = \$63.50$
Gross wage = $444.50 + 76.20 + 63.50 = \$584.20$

EXERCISE 5C

1. Peter the trainee electrician works for $24.50 per hour. His overtime rate is time-and-a-half. What does he get paid for 4 hours overtime?

$24.50 \times 4 \times 1.5 = \$147$

2. Jimmy works for $18.50 per hour. His overtime rate is "double time". What does he get paid for 5 hours overtime?

$18.50 \times 5 \times 2 = \$185$

3. A machinist is paid $21.60 per hour during normal working hours (9 am – 4 pm). For each hour after 4 pm, he is paid time-and-a-half. Calculate the amount received by the machinist on a day when he worked 9am until 6 pm.

$21.60 \times 7 = \$151.20$
$21.60 \times 2 \times 1.5 = \$64.80$
Amount = $151.20 + 64.80 = \$216$

4. Alex is paid double time for each hour he works over 36 hours a week. Last week he worked 41 hours and earned $533.60. Calculate his normal hourly rate.

He worked 5 hours overtime
This is equivalent to
$5 \times 2 = 10$ *hours* of normal hourly rate work
He worked $36 + 10 = 46$ *hours* at the normal hourly rate
Normal hourly rate = $\frac{533.60}{46} = \$11.60$

5. Calculate how much Thomas earns if he works 25 hours at the normal rate of $16.20 per hour, 3 hours at time-and-a-half, and 4 hours at double time.

$16.20 \times 25 = \$405$
$16.20 \times 3 \times 1.5 = \$72.90$
$16.20 \times 4 \times 2 = \$129.60$
Total = $\$607.50$

6. In a week, Sammy works 15 hours at the normal rate, 4 hours at time-and-a-half, and 2 hours at double time. What is his hourly rate if he earns $285?

Total hours worked
$= 15 + 4 \times 1.5 + 2 \times 2 = 25$ hours
Hourly rate = $285 \div 25 = \$11.40$

7. Harry is paid $18 per hour for a 38 hour week. If he works 45 hours in a week with overtime being paid time-and-a-half, how much does he earn for the week?

$18 \times 38 = \$684$
$18 \times 7 \times 1.5 = \$189$
He earns = $684 + 189 = \$873$

8. Gary is paid $16.50 per hour for a 40 hour week. If he works 51 hours in a week with 5 hours being paid time-and-a-half and the rest double-time, how much does he earn for the week?

$16.50 \times 40 = \$660$
$16.50 \times 5 \times 1.5 = \$123.75$
$16.50 \times 6 \times 2 = \$198$
Total = $\$981.75$

5D PIECEWORK

Piecework refers to any type of job where a worker is paid a fixed rate for each unit produced irrespective of time taken. Many companies use the piecewise strategy to encourage efficiency and to increase productivity. However, there are other costs associated with piecework such as Quality Control.

EXAMPLES

1. Jay's Construction Company makes sandbags. Jay gets paid $0.32 a bag. He makes 158 bags a day, six days a week. What is his gross pay?

SOLUTION

Gross pay = $158 \times 6 \times 0.32 = \$303.36$

2. Peter works five days a week for a plastic company that pays him for each unit he completes. He receives 70 cents a unit for the first 80 units he produces each day and 75 cents a unit for all units over 80. Find his weekly income if he produces 105 units on average daily.

SOLUTION

Daily income
$= 80 \times 0.70 + (105 - 80) \times 0.75$
$= \$74.75$
Weekly income $= 5 \times 74.75 = \$373.75$

EXERCISE 5D

1. Peppa Sweet Company makes lolly bags. George gets paid $0.18 a bag. He makes 244 bags a day, five days a week. What is George's gross pay?

$244 \times 5 \times 0.18 = \$219.60$

2. Samantha has a part-time job at Christmas, working for Australia Post. She receives $2.50 for each package she delivers. She delivers 24 packages on average in an evening. What will her gross pay be if she works four nights a week?

$24 \times 4 \times 2.50 = \$240$

3. Gary works six days a week for a garment company that pays him for the units he completes. He receives 75 cents a unit for the first 60 units he produces each day, 80 cents per unit for the next 40, and 90 cents a unit for all units over 100. Find his weekly income if he produces 110 units on average daily.

$60 \times 0.75 + 40 \times 0.80 + 10 \times 0.90 = \$86$
Weekly income $= 86 \times 6 = \$516$

4. Sarah makes socks at the local factory. She is paid 65 cents for every pair of socks she makes. If she made 1672 pair of socks this week, what is Sarah's gross pay?

$1672 \times 0.65 = \$1086.80$

5. Catherine hand paints wall plaques. She is paid $0.85 for every plaque she paints. Catherine painted 495 plaques this week. What is her gross pay?

$495 \times 0.85 = \$420.75$

6. Kat delivers magazines for a local restaurant. She is paid $0.32 per magazine she delivers. Kat is paid weekly. If she delivered 1452 magazines this week, what is her gross pay?

$1452 \times 0.32 = \$464.64$

7. Terri works on her parents grape vines. She is paid $30 plus $0.85 for each basket she fills. If she managed to fill 65 baskets, how much did she earn?

$65 \times 0.85 + 30 = \$85.25$

8. Anthony delivers phone books in his suburb. He is paid $16.25 plus 23 cents for each phone book. If he delivered 280 phone books, how much did he make?

$280 \times 0.23 + 16.25 = \$80.65$

9. Leonardo is a finisher in a furniture plant. He is paid $1.40 for every coffee table he finishes and $1.05 for every end table he finishes. Leonardo finished 198 coffee tables and 395 end tables this week. What is Leonardo's gross pay for the week?

$198 \times 1.40 + 395 \times 1.05$
$= \$691.95$

10. Alexa assembles porcelain dolls at a factory. She is responsible for putting on the legs. She is paid $32 per day, plus 18 cents for each pair of legs she attaches. Yesterday, she attached 398 pairs of legs. How much did she earn?

$32 + 398 \times 0.18 = \$103.64$

11. Joshua is a cheese packer at Long River Cheese Factory. He works every day except Saturday. He is paid $0.48 for every case of cheese he packs and loads onto delivery pallets Monday through Friday and $0.63 for every case he packs and loads on Sunday. Joshua packed and loaded the following cases of cheese this week:

Monday	Tuesday	Wednesday	Thursday	Friday	Sunday
235	245	210	285	190	312

What is Joshua's gross pay for the week?

$235 + 245 + 210 + 285 + 190 = 1165$

Pay foe the week $= 1165 \times 0.48 + 312 \times 0.63 = \$755.76$

CHAPTER 5 : FINANCIAL MATHEMATICS SOLUTIONS

5E   EXCHANGE RATES

The rate of exchange between different currencies varies on a daily basis depending on lot of factors such as economic stability, political situation, social factors and so on.

The table below shows the exchange rates of one Australian dollar (AUD) vis-à-vis other currencies as at 10th July 2014.

TABLE 1

Currency	Symbol	Exchange Rate
US Dollar (USD)	$	0.94
British Pound Sterling (GBP)	£	0.55
Chinese Yuan (CNY)	¥	5.82
Euro (EUR)	€	0.69
Hong Kong Dollar (HKD)	HK$	7.26
Japanese Yen (JPY)	¥	95.11
New Zealand Dollar (NZD)	$	1.06
South African Rand (ZAR)	R	10.05
Swiss Franc (CHF)	CHF	0.84
Indonesian Rupiah (IDR)	Rp	10841
Thai Baht (THB)	฿	30.17

EXAMPLES

1. Matthew changes 25 200 Japanese Yen into Australian dollars.
Calculate how many dollars he receives.
SOLUTION
$\frac{25200}{95.11} = \$264.96$

2. Alex exchanges New Zealand Dollars into Australian Dollar. Bob exchanges British Pound Sterling into Australian Dollar as well. Given that both received $250 (AUD) in return for their exchanges, determine how many NZD did Alex change and the number of GBP Bob converted.
SOLUTION
Alex : $250 \times 1.06 = 265$ NZD

Bob : $250 \times 0.55 = 137.50$ GBP

3. Eddie travels from Australia to Hong Kong. He changes $450 into Hong Kong Dollar.
   (a) How many Hong Kong Dollar does he receive?

   $450 \times 7.26 = \$3267$

   (b) When Eddie returns to Australia he has 81 HKD. How many $ AUD does he get in return?

   $\frac{81}{7.26} = \$11.16$

EXERCISE 5E

1. Complete the following table.

Currency	Exchange Rate to $AUD	Amount converted	Australian Dollars received
US Dollar (USD)	0.94	320	$320 \div 0.94 = \$340.43$
British Pound Sterling (GBP)	0.55	160	$290.91
Chinese Yuan (CNY)	5.82	4200	$721.65
Euro (EUR)	0.69	240	$347.83
Hong Kong Dollar (HKD)	7.26	600	$82.64
Japanese Yen (JPY)	95.11	45 000	$473.14
New Zealand Dollar (NZD)	1.06	680	$641.51
South African Rand (ZAR)	10.05	860	$85.57
Swiss Franc (CHF)	0.84	450	$535.71
Indonesian Rupiah (IDR)	10841	2 000 000	$184.48
Thai Baht (THB)	30.17	60 000	$1988.73

2. Convert the following Australian dollars to the different currencies.

Amount of Australian dollars converted	Currency converted to	Exchange Rate	Amount received
4600	US Dollar (USD)	0.94	$4600 \times 0.94 = \$4324$
150	British Pound Sterling (GBP)	0.55	£ 82.50
5700	Chinese Yuan (CNY)	5.82	¥ 33174
320	Euro (EUR)	0.69	€220.80
800	Hong Kong Dollar (HKD)	7.26	HK$ 5808
700	Japanese Yen (JPY)	95.11	¥66577
680	New Zealand Dollar (NZD)	1.06	$720.80
860	South African Rand (ZAR)	10.05	R 8643

CHAPTER 5 : FINANCIAL MATHEMATICS SOLUTIONS

3. On one particular day the rate of exchange between pounds (£) and Australian dollars ($) was £1 = $1.85.
   Calculate
   (a) the number of dollars received in exchange for £150.

   $$150 \times 1.85 = \$277.50$$

   (b) the number of pounds received in exchange for $264.

   $$264 \div 1.85 = £142.70$$

   (c) John buys 24 postcards for £1.30 each. Calculate the total cost, in dollars, of the postcards.

   $$24 \times 1.30 = £31.20$$
   $$31.20 \times 1.85 = \$57.72$$

4. The exchange rate between pounds and Australian dollar is £1 = $1.76.
   Anna converts $280 into pounds.
   Calculate the number of pounds Anna receives.

   $$280 \div 1.76 = \$159.09$$

5. The exchange rate between Australian dollars and euros is $1 = €0.85.
   Ben changes $260 into euros.
   Calculate the number of euros Ben receives.

   $$260 \times 0.85 = \$221$$

6. Five items bought at Freddo's Supermarket are shown on the receipt.

Apples	
Roll	135
Mineral water	1.20
Cheese	1.64
Tomatoes	1.20
Total $	5.90

   The part showing the cost of the apples is missing.

   (a) How much did the apples cost?

   $0.51

   (b) The total cost of $5.90 when converted to euros is €4.80.
   Determine the exchange rate that will enable you to convert dollars ($) to euros (€).

   $$4.80 \div 5.90 = 0.81$$
   $$\therefore \$1 = €0.81$$

   (c) Use your answer to (b) to estimate the cost of cheese in euros.

   $$1.64 \times 0.81 = €1.32$$

5F BEST BUYS

CLASS ACTIVITY

1. Complete the following table

	Cost per gram	Cost per 100g	Cost per kg
$4 for 50g	$4 \div 50 = \$0.08$	$0.08 \times 100 = \$8$	$0.08 \times 1000 = \$80$
$15.30 for 250g	$0.0612	$0.0612 \times 100 = \$6.12$	$61.20
$12.60 for 400g	$0.0315	$3.15	$31.50
$24.60 for 1 kg	$0.0246	$2.46	$24.60
$60 for 2.5kg	$60 \div 2500 = \$0.024$	$2.40	$24

2. Two shops, Food Mart and Jim's Store, both sell Sweet Yoghurts.

Seven Eleven	P Mart
4 for $5.80	5 for $7

   At which shop are Sweet Yoghurts the better value for money? Show all your working.

   Seven Eleven : $5.80 \div 4 = \$1.45$ per cup
   P Mart : $7 \div 5 = \$1.40$ per cup
   P Mart is cheaper

3. Tomatoes cost $16.50 for a 12.5 kg bag at a farm shop. The same type of tomatoes cost $3.75 for a 3 kg bag at a supermarket. Where are the tomatoes the better value, at the farm shop or at the supermarket?

   Farm Shop
   $16.50 \div 12.5 = \$1.32$ per kg
   Supermarket
   $3.75 \div 3 = \$1.25$ per kg
   Supermarket is cheaper

4. Sprite is on offer at Moles and Foodworth.
   Moles : 2 litre bottles on offer 3 for $9.30.
   Foodworth: 24 cans x 330ml on offer for $24.35. Where is Sprite on a better deal?

   Moles: $9.30 \div 6000 = \$0.00155$ per ml
   Foodworth: $24.35 \div (24 \times 330)$
   $= 0.00307$ per ml
   Moles is cheaper.

5G BUDGETING

Budgeting is often described as an estimate of expected income and expenditures for a given period in the future. It is sometimes itemised and the goal of budgeting is to set a total sum of money aside needed for a purpose: planning a holiday during the school breaks. Budgeting also refers to subsist on or live within a particular budget.

FIXED EXPENDITURE V/S DISCRETIONARY SPENDING

A fixed expense is a cost or obligation that occurs regularly and doesn't vary in amount. Examples of Fixed Expenditures can be Rent or Insurance.

A discretionary spending, on the other hand, is an amount of a person's income that is left for spending after personal necessities such as food, shelter, and clothing have been paid.

CLASS ACTIVITY

Categorise the expenses in the table below as fixed or discretionary.

Description	Fixed	Discretionary	Description	Fixed	Discretionary
Mortgage	✓		Groceries	✓	
Restaurants		✓	Music Lesson fee		✓
Vacation		✓	Utilities	✓	
Massage		✓	Car payment	✓	
Gym Subscription		✓	Pets		✓

CLASS ACTIVITY 2

Make a list of the income you receive and the expenses you have each month and prepare a budget for yourself. If your income and expenses are equal, you aren't saving anything. Do you think this is a problem? If your income exceeds your expenditure, do you find it possible to save to undertake a project such as buying a car, travel etc.. Explain your thinking.

Discuss with classroom teacher

EXERCISE 5G

1. The following table shows the income and expenditure for the Williams family for one year.

Income ($)		Expenditure ($)	
Salaries & Wages	28 240.00	Food	5400.00
Centre Link	6 350.64	Rent	9 600.00
		Power	3 453.60
		Insurance	950.55
		Car payment	1 456.14
		Other	6523.41
		Entertainment	4523.69
Total	**34590.64**	**Total**	**31907.39**

(a) Complete the table by calculating the total income and expenditure.

(b) What percentage of their income did the Williams family save?

$$\frac{34590.64 - 31907.39}{34590.64} \times 100 = 7.76\%$$

(c) The Williams family is planning for a holiday costing $12000. If they plan to set aside 75% of the savings for their holiday, how long would it take for the family to attain their goal?

$$savings = 34590.64 - 31907.39 = 2683.25$$
$$2683.25 \times 0.75 = 2012.44$$
$$Time\ to\ attain\ goal = 12000 \div 2012.44 = 5.96 \sim 6\ years$$

(d) Split the expenditure items under the headings fixed expenditure or discretionary spending.

Fixed expenditure: Rent, Insurance, Car payment

Discretionary spending: Food, Power, Other, Entertainment

CHAPTER 5 : FINANCIAL MATHEMATICS SOLUTIONS

2. Jack and Kelly are both 14 years of age attending the same school. Their parents have agreed they could purchase a car in three years when they both get their licenses at 17. Their aim is to save enough over the next three years to buy the car themselves. They also have to pay for costs like gas, repairs, and insurance. Jack and Kelly found a reasonably priced car for $5000, an amount that they thought they could afford. Bearing inflation in mind the car might cost $5500 in 3 years' time. They decided to make a budget estimate of their expected income and expenditures. Their budget is tabulated below for each month.

Monthly Income & Expenditure	Jack	Kelly
Allowance	$75	$75
Games rental	$8	$0
Part-time job	$128	$116
Snacks	$21	$13
School supplies	$11	$19
Phone rental	$15	$15
Entertainment	$22	$15

(a) List the income and expenditure for both Jack and Kelly and calculate how much they save each month.

Income : Allowance and Part-time job for both
Expenditure for Jack: Games rental, Snacks, School Supplies, Phone rental and Entertainment
Expenditure for Kelly: Snacks, School Supplies, Phone rental and Entertainment
Jack saves per month $(75 + 128) - (8 + 21 + 11 + 15 + 22) = \$126$
Kelly saves per month $(75 + 116) - (13 + 19 + 15 + 15) = \$129$

(b) After three years, will they have saved enough to afford the car?

Jack saves $126 \times 36 = \$4536$
Kelly saves $129 \times 36 = \$4644.$
No. Both falls short of their target.

(c) How much more money does each one need to save each month to afford to buy the car?

Jack needs to save $(5500 - 4536) \div 36 = \$26.78$ **more per month**
Kelly needs to save $(5500 - 4644) \div 36 = \$23.78$ **more per month**

(d) Jack and Kelly are budgeting to make sure they save enough to buy the car. They also have to consider the expenses they will face to operate the car after they buy it. List a few operating expenses they might include.

Gas, repairs, Insurance etc…

3. The table below shows the charges associated with water usage in a Royal City.

Rates for reading the water meter	
Usage (kL) per year	Meters read January–December 2014
first 150 kL	85.2 c/kL
Next 200 kL	120.5 c/kL
next 150 kL	132.6 c/kL
over 500 kL	169.4 c/kL

(a) The account shows that the water usage for the Smith family was 230 kL. Calculate the amount they are required to pay for their water usage.

$150 \times 85.2 = 12780c$
$80 \times 120.5 = 9640c$
$12780 + 9640 = 22420c = \$224.20$

(b) The Bligh's water usage was 450 kL. Calculate the amount they are required to pay for their water usage.

$150 \times 85.2 = 12780c$
$200 \times 120.5 = 24100c$
$100 \times 132.6 = 13260$
$12780 + 24100 + 13260 = 50140c = \$501.40$

(c) The Simpson's have a budget of $500 for their water usage. Calculate by how much they exceeded their budget if their water consumption was exactly 540 kL.?

$150 \times 85.2 = 12780c$
$200 \times 120.5 = 24100c$
$150 \times 132.6 = 19890$
$40 \times 169.4 = 6776$
$12780 + 24100 + 19890 + 6776 = 63546c = \$635.46$

Budget exceeded by $635.46 - 500 = \$135.46$

CHAPTER 5 : FINANCIAL MATHEMATICS SOLUTIONS

5H SHARES AND DIVIDENDS

Dividend is the distribution of a fraction of a company's earnings to its shareholders, usually decided by the board of directors. Dividend is often quoted in terms of the dollar amount each share receives also referred to as "Dividend Per Share (DPS)."

Dividends may be in the form of cash, stock or property. Most secure and stable companies offer dividends to their shareholders. High-growth companies rarely offer dividends because all of their profits are reinvested to help sustain higher-than-average growth.

EXAMPLE 1

Calculate the dividend paid on the portfolio of shares given the dividend paid for each share.

Company	Number of shares	Price per share	Dividend per share
AB Co Ltd	2000	$1.20	$0.15
DNDS Inc	450	$4.60	$0.60
Golden Miners	6000	$36.55	$2.40

SOLUTION

Dividend paid by AB Co Ltd = $2000 \times 0.15 = \$300$

Dividend paid by DNDS Inc = $450 \times 0.60 = \$270$

Dividend paid by Golden Miners = $6000 \times 2.40 = \$14400$

Total dividend = $14970

EXAMPLE 2

Calculate the dividend paid on the portfolio of shares given the percentage dividend paid per share.

Company	Number of shares	Price per share	Percentage Dividend per share
Junior Ltd	5500	$8.60	5.6%
The Croods Inc	680	$15.40	3.9%
Philip Trading	4250	$3.75	No dividend

SOLUTION

Dividend paid by Junior Ltd = $5500 \times 8.60 \times 5.6\% = \$2648.80$

Dividend paid by Croods Inc = $680 \times 15.40 \times 3.9\% = \$408.41$

Dividend paid by Philip Trading = $4250 \times 0 = \$0$

Total dividend = $3057.21

EXERCISE 5H

1. Calculate the dividend paid on the portfolio of shares given the dividend paid for each share.

Company	Number of shares	Price per share	Dividend per share
Avengers Ltd	800	$2.20	$0.25
Knight Inc	1250	$10.60	No Dividend
Thor Mining	630	$15.40	$1.70
Green lantern Ltd	5800	$5.80	$0.90

Dividend paid by Avengers Ltd = $800 \times 0.25 = \$200$

Dividend paid by Knight Inc = $1250 \times 0 = \$0$

Dividend paid by Thor Mining = $630 \times 1.70 = \$1071$

Dividend paid by Green Lantern Ltd = $5800 \times 0.90 = \$5220$

Total dividend = $6491

2. Calculate the dividend paid on the portfolio of shares given the percentage dividend paid per share.

Company	Number of shares	Price per share	Percentage Dividend per share
Alpha Ltd	650	$12.40	4.6%
Beta Corp	7000	$7.20	5.1%
Gamma Inc	980	$11.55	No dividend
Delta Mining	2500	$6.90	8.3%

Dividend paid by Alpha Ltd = $650 \times 12.40 \times 4.6\% = \$370.76$

Dividend paid by Beta Corp = $7000 \times 7.20 \times 5.1\% = \$2570.40$

Dividend paid by Gamma Inc = $0

Dividend paid by Delta Mining = $2500 \times 6.90 \times 8.3\% = \$1431.75$

Total dividend = $4372.91

CHAPTER 5 : FINANCIAL MATHEMATICS SOLUTIONS

3. Calculate the dividend paid on the portfolio of shares given the dividend paid for each share.

Company	Number of shares	Price per share	Dividend per share
Company A	3200	$2.90	$0.35
Company B	780	$11.35	$1.30
Company C	5120	$8.20	No dividend
Company D	2670	$6.40	$1.42
Company E	925	$16.60	$2.65

Dividend paid by Company A = $3200 \times 0.35 = \$1120$

Dividend paid by Company B = $780 \times 1.30 = \$1014$

Dividend paid by Company C = $5120 \times 0 = \$0$

Dividend paid by Company D = $2670 \times 1.42 = \$3791.40$

Dividend paid by Company E = $925 \times 2.65 = \$2451.25$

Total dividend = $8376.65

4. Calculate the dividend paid on the portfolio of shares given the percentage dividend paid per share.

Company	Number of shares	Price per share	Percentage Dividend per share
Murray Ltd	1230	$10.20	2.1%
Swan Corp	3500	$5.30	4.6%
Canning Inc	2800	$9.75	5.3%

Dividend paid by Murray Ltd = $1230 \times 10.20 \times 2.1\% = \$263.47$

Dividend paid by Swan Corp = $3500 \times 5.30 \times 4.6\% = \$853.30$

Dividend paid by Canning Inc = $2800 \times 9.75 \times 5.3\% = \$1446.90$

Total dividend = $2563.67

51 PRICE–TO–EARNINGS RATIO

Price-to-earnings ratio, commonly known as P/E ratio, is a tool that is used by investors deciding whether they should buy a particular stock or not. The P/E ratio implies how much an investor has to pay for every $1 of earnings. A low P/E ratio is attractive implying that one pays less for every $1 of earnings. Similarly, a company with a higher P/E ratio generally expect higher earnings growth in the future.

$$Price-to-Earnings\ ratio = \frac{Price\ per\ share}{Earnings\ per\ share}$$

Price per share in simple words is the cost of buying a share of any company on the stock Market. Earnings per share, on the other hand, is calculated by taking a company's net income over the last twelve months, subtracting any dividends, and then dividing the difference by the number of shareholders.

CLASS ACTIVITY

1. Calculate the P/E ratio for each of the following companies.

Company	Market value per share	Earnings per share	P/E Ratio
DK Trading	15.50	1.50	$\frac{15.50}{1.50} = 10.3$
RJ Mining	8.60	0.80	**10.75**
ABC Banking	1.65	0.25	**6.6**
Mars Bar Co Ltd	49.43	2.55	**19.38**
Top Shop Ltd	12.60	2.40	**5.25**

2. Complete the following table.

Company	Market value per share	Earnings per share	P/E Ratio
SOS Inc	18.60	**2.40**	7.75
Delta Trading	9.25	1.25	7.4
Swiss Choc Factory	15.60	**1.50**	10.4
MBCL Ltd	**22.80**	1.20	19
Alien Groups	2.80	0.40	7

3. Alan has the choice to buy shares in two different companies A and B. He collected some information about both companies as shown under.

   Company A   Company B
   Price per share : $16   Price per share : $15
   EPS : $2   EPS: $1
   P/E ratio = 8   P/E Ratio = 15

   Which company do you advise him to invest in?

   SOLUTION

   Company A has a P/E ratio of 8 which means Alan will be paying $8 to earn $1.
   Company B, on the other hand, has a P/E ratio of 15 implying that Alan needs to pay $15 for every dollar of earning.
   Company A having a low P/E ratio is better as we are paying less money for a company that earns more.

   Although a low P/E ratio is a great indicator that Company A could be selling for an attractive price. However, it is not the only indicator. Alan must make sure he uses other measures and make an informed decision before investing in either company.

4. You have been assigned to evaluate the following three companies stocks within the same industry:

Company	Price per share	Earnings Per Share
Cybertrons Corporation	$52.40	$12.50
Decepticons Ltd	$8.50	$2.25
Optimus Prime Inc	$9.88	$5.25

   Which stock has the best value? Show workings.

   SOLUTION

   P/E ratio for Cybertrons Corp = $\frac{52.40}{12.50}$ = **4.2**

   P/E ratio for Decepticons Ltd = $\frac{8.50}{2.25}$ = **3.8**

   P/E ratio for Optimus Prime Inc = $\frac{9.88}{5.25}$ = **1.9**

   All the three companies being in the same industry, Optimus Prime having the lowest P/E ratio will be the best option.

5J   ALLOWANCES : FAMILY TAX BENEFIT

Family Tax Benefit (FTB) is an allowance that helps eligible families with the cost of raising their children. It is made up of two parts FTB Part A and FTB Part B. FTB Part A is paid depending on the number of children in a family and their financial status. FTB Part B, on the other hand, is paid per-family and gives extra help to single parents and families with one main income.

The maximum amounts of Family Tax Benefit Part A received per child are updated on 1 July each year. The table below shows the current rates per child within different age brackets and circumstances. For the scope of this book, only FTB part A has been included as examples.

TABLE 1 : CURRENT RATES FTB Part A

For each child	Per fortnight
0 to 12 years	$169.68
13-15 years	$220.64
16-19 years	$220.64
16-17 years (completed secondary study)	$54.32
18-21 years (completed secondary study)	$54.32

To be eligible for the above rates, a particular family must undergo one of the following tests depending on their yearly income.

TABLE 2 : FTB Part A Tests

TEST 1	Maximum rate for Family Tax Benefit Part A less 20 cents for each dollar above $50,151.
TEST 2	Maximum rate for Family Tax Benefit Part A less 30 cents for each dollar above $94,316, plus $3,796 for each Family Tax Benefit child after the first.

EXAMPLES

1. The Smith family has two children aged 5 and 8 years. The combined family income is $63500. Calculate the amount of FTB Part A received by the Smith family.
   SOLUTION
   Yearly maximum amount for both children
   = 169.68 × 2 × 26 = $8823.36
   Applying the first test :
   (63500 − 50151) × 0.20 = $2669.80
   FTB Part A received
   8823.36 − 2669.80 = $6153.56

2. The Gordon family has two children aged 10 and 17 years. The 17 year old son has completed his secondary study and works part-time in a restaurant. The family net income is $59860. Determine the amount of FTB Part A received by the Gordon family.
   SOLUTION
   Yearly maximum amount for both children
   = 169.68 × 26 + 54.32 × 26 = $5824
   Applying the first test :
   (59860 − 50151) × 0.20 = $1941.80
   FTB Part A received
   5824 − 1941.80 = $3882.20

CHAPTER 5 : FINANCIAL MATHEMATICS SOLUTIONS

EXERCISE 5J

Referring to Tables 1 and 2 on the previous page, answer the following questions.

1. A family's income is $55765 a year. The family has two children of 3 and 9 years respectively. Calculate the amount of FTB Part A received by this family.

Yearly maximum amount for both children
$= 169.68 \times 2 \times 26 = \$8823.36$
Applying the first test :
$(55765 - 50151) \times 0.20 = \$1122.80$
FTB Part A received
$8823.36 - 1122.80 = \$7700.56$

2. Mr and Mrs Pavilion have three children : a son aged 11 years and twin daughters of 15 years of age. The Pavilions' run their own business and earn $75420 a year. Calculate the amount of FTB Part A received by the family.

Yearly maximum amount for three children
$= 169.68 \times 26 + 220.64 \times 2 \times 26 = \$15884.96$
Applying the first test :
$(75420 - 50151) \times 0.20 = \$5053.80$
FTB Part A received
$15884.96 - 5053.80 = \$10831.16$

3. Calculate the amount of FTB Part A received by the Packard family having only one child aged 14 years and having a net annual income of $90000.

Yearly maximum amount for one child
$= 220.64 \times 26 = \$5736.64$
Applying the first test :
$(90000 - 50151) \times 0.20 = \$7969.80$
FTB Part A received
NIL

4. A family's income is $1985 a week. The family has two children of 7 and 16 years respectively. Calculate the amount of FTB Part A received by this family.

Yearly income $= 1985 \times 52 = \$103220$
Yearly maximum amount for both children
$= (169.68 + 220.64) \times 26 = \$10148.32$
Applying Test 2 :
$(103220 - 94316) \times 0.30 = \$2671.20$
FTB Part A received
$10148.32 - (2671.20 + 3796) = \$3681.12$

5K   CARER ALLOWANCE AND OLD AGE PENSIONS

CARER ALLOWANCE : A carer allowance is an additional payment made to people who provide additional daily care and attention for someone with a disability or medical condition. Carer Allowance when caring for a child less than 16 years is currently a fortnightly payment of $118.20. The payments are however adjusted on 1st January each year to match the cost of living. Note that Carer Allowance is a non-taxable payment.

AGE PENSIONS : In order to be eligible for Age Pensions in Australia, a person has to meet the residence requirements. If you are an Australian citizen and have been residing in Australia for at least ten years cumulatively then you can be eligible for the age pension. Permanent residents living in Australia may also be eligible for age pension.

In addition to residence requirements, there's also an assets test and income test the Government will carry out to assess whether a particular person is eligible for the Age Pension or not. The tests take into account factors such as age pension, wife pension, carer payment and so on.

The table below shows the Pension rates for Age Pensions.

TABLE 1

Pension rates for Age Pensions		
Pension rates per fortnight	Single	Couple each
Maximum basic rate	$776.70	$585.50

The **ASSETS TEST** has two thresholds and is split into two categories as shown in the table below.

TABLE 2
THE ASSETS TEST

	Home Owners		Non-Home Owners	
	Lower Threshold	Upper threshold	Lower Threshold	Upper threshold
Single	192 500	707 750	332 000	847 250
Couples	273 000	1 050 000	412 500	1 189 500

The full age pension is received when a lower assets test threshold is not exceeded. Once the lower thresholds are exceeded a person or couple's entitlement to the age pension is reduced by $1.50 a fortnight for every $1000 their assets exceed that threshold. Unfortunately, no age pension is received once an upper threshold is exceeded.
The table below shows the **INCOME TEST** for singles and couples.

TABLE 3
THE INCOME TEST

	Payment per fortnight	Reductions
Single	Up to $160	None – full payment
	Above $160	50 cents for each dollar over $160
Couples combined	Up to $284	None – full payment
	Above $284	50 cents for each dollar over $284

For a person being single and earning income up to $160 a fortnight, there is no deduction in his or her pension. However, for income exceeding $160, there is a deduction of 50 cents per dollar.

EXAMPLES

1. Using the information from Table 1 and Table 2 on the previous page, determine the fortnightly age pension of a single non-homeowner of pension age owning assets worth $390 000.

SOLUTION

Fortnightly Age pension

$$= 776.70 - \left(\frac{390000 - 332000}{1000}\right) \times 1.50$$
$$= \$689.70$$

2. Mr and Mrs Jones, both in their late sixties, have invested a sum of $200 000 which earns them a yearly simple interest of 5%. Calculate their combined fortnightly age pension using Table 1 and Table 3.

SOLUTION

$Interest\ earned = 200000 \times 0.05$
$= \$10000$

Fortnightly income $= 10000 \div 26 \approx \$385$

Fortnightly Age pension
$= 585.50 \times 2 - (385 - 284) \times 0.50$
$= \$1120.50$

EXERCISE 5K

1. Using the information from Table 1 and Table 2 on the previous page, determine the fortnightly age pension of a single homeowner of pension age owning assets worth $405 000.

$$776.70 - \left(\frac{405000 - 192500}{1000}\right) \times 1.50$$
$$= \$457.95$$

2. Mr Adan Bligh, a widower in his early seventies, has invested a sum of $150 000 which earns him a flat fixed rate of 6% each year. Calculate his fortnightly age pension using Table 1 and Table 3.

$Interest\ earned = 150000 \times 0.06$
$= \$9000$

Fortnightly income
$= 9000 \div 26 \approx \$346$

Fortnightly Age pension
$= 776.70 - (346 - 160) \times 0.50$
$= \$683.70$

3. Using the information from Table 1 and Table 2 on the previous page, determine the fortnightly age pension of a home owning couple both of pension age owning assets worth $580 000.

$$585.50 \times 2 - \left(\frac{580000 - 273000}{1000}\right) \times 1.50$$
$$= \$710.50$$

4. Mrs Blitz is a 69 year old widow. She has invested $90 000 as a shareholder in an IT company. Last year she received 2% flat interest on her investment plus $450 as dividend each fortnight. Using Tables 1 and 3, calculate her fortnightly age pension.

$Interest\ earned = 90000 \times 0.02$
$= \$1800$

Fortnightly income
$= 1800 \div 26 + 450 \approx \$519$

Fortnightly Age pension
$= 776.70 - (519 - 160) \times 0.50$
$= \$579.20$

5. Agent Coulson is a single homeowner and a retired army officer and has reached the pension age. He has assets worth $365 000. He has invested $165 000 of his assets in Delta Bank earning 4% per annum as interest.

(a) Determine using the Assets Test his fortnightly pension.

$$776.70 - \left(\frac{365000 - 192500}{1000}\right) \times 1.50$$
$$= \$571.95$$

(b) Determine using the Income Test his fortnightly age pension.

$Interest\ earned = 165000 \times 0.04$
$= \$6600$

Fortnightly income
$= 6600 \div 26 \approx \$254$

Fortnightly Age pension $= 776.70 - (254 - 160) \times 0.50$
$= \$729.70$

(c) It is customary for the government to pay the lower pension out of the Income test or Assets test carried out. State Agent Coulson's fortnightly pension rounded up to the nearest dollar.

$572

CHAPTER 6

MATRICES

A matrix is an array of numbers arranged in rows (horizontally) or in columns (vertically) or both. Consider the following example.

A café sells tea, juice and coffee, each in small and large cups.
The cost of a small cup is $3.50 for all three types of drinks and the cost for large cup is $4.25.
The table below shows the number of cups sold during a period of one hour.

Drink	Small	Large
Tea	9	5
Juice	12	3
Coffee	7	8

The above table can be represented in a matrix as shown.

$$\begin{bmatrix} 9 & 5 \\ 12 & 3 \\ 7 & 8 \end{bmatrix} \rightarrow three\ rows$$

2 columns

The numbers represented horizontally are the rows and the numbers written vertically are called columns. In the above matrix there 3 rows and 2 columns. Each of the numbers in the matrix are called the elements of the matrix and are denoted as e_{ij}, i being the row number and j being the column number. For example, $e_{22} = 3$.

As a convention, we use capital letters and square brackets to represent a matrix.

CLASS ACTIVITY 1

$\begin{bmatrix} 4 & 2 & 5 & 8 \\ -1 & 0 & 3 & -7 \\ 3 & 7 & 9 & 1 \end{bmatrix}$	$e_{21} = -1$	$e_{21} = 1$
	$e_{33} = 9$	$e_{32} = 4$
	$e_{14} = 8$	$e_{42} = 6$
$\begin{bmatrix} 5 & 1 & -4 \\ 0 & 8 & 6 \\ 4 & 2 & 7 \end{bmatrix}$	$e_{31} = 4$	$\begin{bmatrix} 5 & -2 & -5 \\ 1 & 3 & -3 \\ 2 & 4 & 0 \\ 7 & 8 & 9 \end{bmatrix}$	$e_{23} = -3$
	$e_{23} = 6$		$e_{32} = 4$
	$e_{12} = 1$		$e_{41} = 7$

6A   SIZE (ORDER) OF A MATRIX

Consider the matrix $\begin{bmatrix} 9 & 5 \\ 12 & 3 \\ 7 & 8 \end{bmatrix}$. Since it has three rows and two columns, we say the order of the matrix is 3×2 (read as 3 by 2). Therefore,

Order of a matrix = number of rows (r) × number of columns (c).

CLASS ACTIVITY 2

Complete the following table, where necessary.

MATRIX	SIZE	MATRIX	SIZE
$\begin{bmatrix} 5 & 2 & -1 \\ 6 & 4 & 7 \end{bmatrix}$	2×3	$\begin{bmatrix} 6 \\ -2 \\ 5 \end{bmatrix}$	3×1
$\begin{bmatrix} 1 & 7 \\ 5 & 4 \end{bmatrix}$	2×2	$\begin{bmatrix} 4 & 8 \\ 0 & 1 \\ 7 & 3 \end{bmatrix}$	3×2
$[7 \ 11]$	1×2	$\begin{bmatrix} 4 & 0 \\ 1 & 2 \\ 3 & 4 \\ 5 & 6 \end{bmatrix}$	4×2
$\begin{bmatrix} 5 \\ -1 \\ 3 \\ 8 \end{bmatrix}$	4×1	$[11]$	1×1
$[0 \ 1 \ 2]$	1×3	$\begin{bmatrix} 5 & 1 & 0 \\ 0 & 8 & 6 \\ 4 & 2 & 7 \end{bmatrix}$	3×3
$\begin{bmatrix} 4 & 2 & 5 & 2 \\ -1 & 0 & 3 & 7 \\ 3 & 7 & 9 & 1 \end{bmatrix}$	3×4	$\begin{bmatrix} 1 \\ 2 \\ 4 \\ 7 \\ 3 \end{bmatrix}$	5×1

6B TYPES OF MATRICES

There are several types of matrices but in this section emphasis will be laid on row, column, square, zero, diagonal, identity and equal matrices.

ROW MATRIX

It is a matrix having only one row. The size of a row matrix is in the form $1 \times c$, where c represents the number of columns.

Examples of row matrices are $P = [0 \ 1 \ 2]$ or $Q = [5 \ 2 \ 4 \ 7]$.

COLUMN MATRIX

It is a matrix having only one column. Since there is only one column the size of a column matrix is always written in the form $r \times 1$, where r represents the number of rows.

Examples of column matrices are : $A = \begin{bmatrix} 6 \\ -2 \\ 5 \end{bmatrix}$ or $B = \begin{bmatrix} 1 \\ 2 \\ 4 \\ 7 \\ 3 \end{bmatrix}$

SQUARE MATRIX

It is matrix in which the number of rows is equal to the number of columns. Examples of square matrices are $C = \begin{bmatrix} 1 & 7 \\ 5 & 4 \end{bmatrix}$ or $D = \begin{bmatrix} 5 & 1 & 0 \\ 0 & 8 & 6 \\ 4 & 2 & 7 \end{bmatrix}$ or $E = [7]$

ZERO MATRIX (NULL MATRIX)

It is a matrix in which all the elements are zeroes.

Examples of zero matrices are $M = \begin{bmatrix} 0 & 0 \\ 0 & 0 \end{bmatrix}$ or $N = \begin{bmatrix} 0 & 0 & 0 \\ 0 & 0 & 0 \\ 0 & 0 & 0 \end{bmatrix}$

DIAGONAL MATRIX

It is a matrix in which all the elements are zeroes with the exception of those elements lying in the leading diagonal; the latter slopes downwards from left to right.

Examples of diagonal matrices are $M = \begin{bmatrix} 1 & 0 \\ 0 & 3 \end{bmatrix}$ or $N = \begin{bmatrix} 4 & 0 & 0 \\ 0 & 5 & 0 \\ 0 & 0 & -2 \end{bmatrix}$

IDENTITY MATRIX (UNIT MATRIX)

It is a diagonal matrix in which all the elements are ones. We use the letter I to denote an identity matrix.

Examples of identity matrices are $M = \begin{bmatrix} 1 & 0 \\ 0 & 1 \end{bmatrix}$ or $N = \begin{bmatrix} 1 & 0 & 0 \\ 0 & 1 & 0 \\ 0 & 0 & 1 \end{bmatrix}$

EQUAL MATRICES

Two matrices are said to be equal if
(a) They have the same size
(b) Their corresponding elements are equal.

For example if $\begin{bmatrix} x & 0 \\ 2 & 3 \end{bmatrix} = \begin{bmatrix} 1 & 0 \\ 2 & y \end{bmatrix}$, then $x = 1 \ and \ y = 3$

CLASS ACTIVITY 3

Name the following matrices.

MATRIX	NAME	MATRIX	NAME
$\begin{bmatrix} 5 & 0 & 0 \\ 0 & 8 & 0 \\ 0 & 0 & 7 \end{bmatrix}$	Square diagonal	$\begin{bmatrix} 6 \\ -2 \\ 5 \end{bmatrix}$	*column*
$\begin{bmatrix} 1 & 7 \\ 5 & 4 \end{bmatrix}$	Square	$\begin{bmatrix} 4 & 8 \\ 0 & 1 \\ 7 & 3 \end{bmatrix}$	None
$[7 \ 11]$	row	$\begin{bmatrix} 4 & 0 \\ 1 & 2 \\ 3 & 4 \\ 5 & 6 \end{bmatrix}$	None
$\begin{bmatrix} 1 & 0 \\ 0 & 1 \end{bmatrix}$	Square Diagonal Identity	$[11]$	**Row Column Square**
$[0 \ 0 \ 0]$	**Row Zero**	$\begin{bmatrix} 0 \\ 0 \end{bmatrix}$	**column zero**
$\begin{bmatrix} 2 & 3 & 4 \\ 1 & -8 & 3 \\ 4 & 0 & -3 \end{bmatrix}$	Square	$\begin{bmatrix} 0 & 1 \\ 1 & 0 \end{bmatrix}$	Square

CHAPTER 6 : MATRICES SOLUTIONS

6C ADDITION OF MATRICES

Matrices can be added provided they have the same size. Adding the corresponding elements in each matrix performs addition.

EXAMPLE

Given $\quad A = \begin{bmatrix} 1 & 7 \\ 5 & 4 \end{bmatrix} \quad B = \begin{bmatrix} 3 & 5 \\ -2 & 0 \end{bmatrix} \quad C = \begin{bmatrix} 10 \\ 15 \end{bmatrix}$,

express as a single matrix.

(a) $A + B$ \quad (b) $B + C$

SOLUTION

(a) $A + B = \begin{bmatrix} 1 & 7 \\ 5 & 4 \end{bmatrix} + \begin{bmatrix} 3 & 5 \\ -2 & 0 \end{bmatrix} = \begin{bmatrix} 4 & 12 \\ 3 & 4 \end{bmatrix}$

(b) $B + C = \begin{bmatrix} 3 & 5 \\ -2 & 0 \end{bmatrix} + \begin{bmatrix} 10 \\ 15 \end{bmatrix} = $ *Not possible as they have different size.*

EXERCISE 6C

Express the following as single matrices.

1.	$\begin{bmatrix} 5 \\ 3 \end{bmatrix} + \begin{bmatrix} 4 \\ -2 \end{bmatrix} = \begin{bmatrix} \mathbf{9} \\ \mathbf{1} \end{bmatrix}$	**2.**	$\begin{bmatrix} 2 & 7 \\ 8 & 0 \end{bmatrix} + \begin{bmatrix} 8 & -1 \\ -2 & 0 \end{bmatrix} = \begin{bmatrix} \mathbf{10} & \mathbf{6} \\ \mathbf{6} & \mathbf{0} \end{bmatrix}$
3.	$\begin{bmatrix} 5 & 1 & 0 \\ 0 & 8 & 6 \\ 4 & 2 & 7 \end{bmatrix} + \begin{bmatrix} 2 & 3 & 4 \\ 1 & -8 & 3 \\ 4 & 0 & -3 \end{bmatrix}$ $= \begin{bmatrix} \mathbf{7} & \mathbf{4} & \mathbf{4} \\ \mathbf{1} & \mathbf{0} & \mathbf{9} \\ \mathbf{8} & \mathbf{2} & \mathbf{4} \end{bmatrix}$	**4.**	$\begin{bmatrix} 6 \\ -2 \\ 5 \end{bmatrix} + \begin{bmatrix} 4 & 8 \\ 0 & 1 \\ 7 & 3 \end{bmatrix}$ **Not possible as they have different sizes.**
5.	$\begin{bmatrix} 4 & 3 \\ 6 & 5 \end{bmatrix} + \begin{bmatrix} 6 & -3 \\ -1 & 8 \end{bmatrix} = \begin{bmatrix} \mathbf{10} & \mathbf{0} \\ \mathbf{5} & \mathbf{13} \end{bmatrix}$	**6.**	$\begin{bmatrix} -6 \\ 3 \end{bmatrix} + \begin{bmatrix} 10 \\ -7 \end{bmatrix} = \begin{bmatrix} \mathbf{4} \\ \mathbf{-4} \end{bmatrix}$
7.	$\begin{bmatrix} 2a & x \\ 8 & 3b \end{bmatrix} + \begin{bmatrix} 8a & 4x \\ 2 & 0 \end{bmatrix} = \begin{bmatrix} \mathbf{10a} & \mathbf{5x} \\ \mathbf{10} & \mathbf{3b} \end{bmatrix}$	**8.**	$\begin{bmatrix} 6 & 2 \\ -2 & 0 \\ 5 & 3 \end{bmatrix} + \begin{bmatrix} 1 & 7 \\ -3 & -5 \\ -4 & 1 \end{bmatrix} = \begin{bmatrix} \mathbf{7} & \mathbf{9} \\ \mathbf{-5} & \mathbf{-5} \\ \mathbf{1} & \mathbf{4} \end{bmatrix}$

MATHEMATICS APPLICATIONS UNIT 1

6D SUBTRACTION OF MATRICES

Similar to addition, matrices can be subtracted provided they have the same size. Subtracting the corresponding elements in each matrix performs subtraction.

EXAMPLE

Given $\quad P = \begin{bmatrix} 4 & -1 \\ 0 & 5 \end{bmatrix} \quad Q = \begin{bmatrix} 3 & 8 \\ -2 & -2 \end{bmatrix} \quad R = \begin{bmatrix} 9 \\ 11 \end{bmatrix}$,

express as a single matrix.

(a) $P - Q$ \quad (b) $Q - R$

SOLUTION

(a) $P - Q = \begin{bmatrix} 4 & -1 \\ 0 & 5 \end{bmatrix} - \begin{bmatrix} 3 & 8 \\ -2 & -2 \end{bmatrix} = \begin{bmatrix} 1 & -9 \\ 2 & 7 \end{bmatrix}$

(b) $Q - R = \begin{bmatrix} 3 & 8 \\ -2 & -2 \end{bmatrix} - \begin{bmatrix} 9 \\ 11 \end{bmatrix} = $ *Not possible as they have different size.*

EXERCISE 6D

Express the following as single matrices.

1.	$\begin{bmatrix} 5 \\ 3 \end{bmatrix} - \begin{bmatrix} 2 \\ -2 \end{bmatrix} = \begin{bmatrix} \mathbf{3} \\ \mathbf{5} \end{bmatrix}$	**2.**	$\begin{bmatrix} 2 & 9 \\ 5 & 0 \end{bmatrix} - \begin{bmatrix} 8 & -1 \\ -2 & 0 \end{bmatrix} = \begin{bmatrix} \mathbf{-6} & \mathbf{10} \\ \mathbf{7} & \mathbf{0} \end{bmatrix}$
3.	$\begin{bmatrix} 5 & 1 & 0 \\ 0 & 8 & 6 \\ 4 & 2 & 7 \end{bmatrix} - \begin{bmatrix} 4 & 8 \\ 0 & 1 \\ 7 & 3 \end{bmatrix}$ **Not possible as they have different sizes.**	**4.**	$\begin{bmatrix} 6 \\ -2 \\ 5 \end{bmatrix} - \begin{bmatrix} 4 & 5 \\ 0 & -2 \\ 7 & 3 \end{bmatrix}$ **Not possible as they have different sizes.**
5.	$\begin{bmatrix} 4 & 3 \\ 6 & 5 \end{bmatrix} - \begin{bmatrix} 6 & 4 \\ -1 & 7 \end{bmatrix} = \begin{bmatrix} \mathbf{-2} & \mathbf{-1} \\ \mathbf{7} & \mathbf{-2} \end{bmatrix}$	**6.**	$\begin{bmatrix} -6 \\ 3 \end{bmatrix} - \begin{bmatrix} 8 \\ -5 \end{bmatrix} = \begin{bmatrix} \mathbf{-14} \\ \mathbf{8} \end{bmatrix}$
7.	$\begin{bmatrix} 12a & 2x \\ 8 & 5b \end{bmatrix} - \begin{bmatrix} 7a & x \\ -2 & 7b \end{bmatrix} = \begin{bmatrix} \mathbf{5a} & \mathbf{x} \\ \mathbf{10} & \mathbf{-2b} \end{bmatrix}$	**8.**	$\begin{bmatrix} 6 & 2 \\ -2 & 0 \\ 5 & 3 \end{bmatrix} - \begin{bmatrix} 1 & 7 \\ -3 & -3 \\ -4 & -4 \end{bmatrix} = \begin{bmatrix} \mathbf{7} \\ \mathbf{1} \\ \mathbf{12} \end{bmatrix}$

CHAPTER 6 : MATRICES SOLUTIONS

6E   MULTIPLYING A MATRIX BY A SCALAR

A scalar is in fact just a random number. To multiply a matrix by a scalar, multiply all the elements of the matrix by the scalar as shown in the examples below.

EXAMPLE 1

Given $A = \begin{bmatrix} 1 & 7 \\ 5 & 4 \end{bmatrix}$ $B = \begin{bmatrix} 3 & 5 \\ -2 & 0 \end{bmatrix}$

Express as a single matrix.

(a) $3A$ (b) $A + 2B$

SOLUTION

(a) $3A = 3\begin{bmatrix} 1 & 7 \\ 5 & 4 \end{bmatrix} = \begin{bmatrix} 3 & 21 \\ 15 & 12 \end{bmatrix}$

(b) $A + 2B = \begin{bmatrix} 1 & 7 \\ 5 & 4 \end{bmatrix} + 2\begin{bmatrix} 3 & 5 \\ -2 & 0 \end{bmatrix}$

$= \begin{bmatrix} 1 & 7 \\ 5 & 4 \end{bmatrix} + \begin{bmatrix} 6 & 10 \\ -4 & 0 \end{bmatrix}$

$= \begin{bmatrix} 7 & 17 \\ 1 & 4 \end{bmatrix}$

EXAMPLE 2

If $3N = \begin{bmatrix} 6 & 12 \\ -3 & 0 \end{bmatrix}$, find N.

SOLUTION

$N = \frac{1}{3}\begin{bmatrix} 6 & 12 \\ -3 & 0 \end{bmatrix} = \begin{bmatrix} 2 & 4 \\ -1 & 0 \end{bmatrix}$

EXAMPLE 3

Find the matrix M which is such that $2M - \begin{bmatrix} 0 & 4 \\ -6 & 8 \end{bmatrix} = 3\begin{bmatrix} 2 & 0 \\ 0 & -4 \end{bmatrix}$

SOLUTION

$2M = \begin{bmatrix} 0 & 4 \\ -6 & 8 \end{bmatrix} + 3\begin{bmatrix} 2 & 0 \\ 0 & -4 \end{bmatrix}$

$= \begin{bmatrix} 0 & 4 \\ -6 & 8 \end{bmatrix} + \begin{bmatrix} 6 & 0 \\ 0 & -12 \end{bmatrix}$

$2M = \begin{bmatrix} 6 & 4 \\ -6 & -4 \end{bmatrix}$

$\therefore M = \begin{bmatrix} 3 & 2 \\ -3 & -2 \end{bmatrix}$

EXERCISE 6E

Express as a single matrix.

1. $3\begin{bmatrix} 4 \\ 5 \end{bmatrix} = \begin{bmatrix} 12 \\ 15 \end{bmatrix}$

2. $2\begin{bmatrix} -3 \\ 8 \end{bmatrix} = \begin{bmatrix} -6 \\ 16 \end{bmatrix}$

3. $3\begin{bmatrix} 3 & 0 \\ 5 & -2 \end{bmatrix} = \begin{bmatrix} 9 & 0 \\ 15 & -6 \end{bmatrix}$

4. $4\begin{bmatrix} 2 & 3 \\ 0.5 & -5 \end{bmatrix} = \begin{bmatrix} 8 & 12 \\ 2 & -20 \end{bmatrix}$

5. $3\begin{bmatrix} 1 \\ 6 \end{bmatrix} + \begin{bmatrix} 7 \\ 2 \end{bmatrix} = \begin{bmatrix} 10 \\ 20 \end{bmatrix}$

6. $2\begin{bmatrix} 5 \\ 4 \end{bmatrix} - \begin{bmatrix} 3 \\ -1 \end{bmatrix} = \begin{bmatrix} 7 \\ 9 \end{bmatrix}$

7. $\frac{1}{2}\begin{bmatrix} 8 & 0 \\ 6 & -2 \end{bmatrix} = \begin{bmatrix} 4 & 0 \\ 3 & -1 \end{bmatrix}$

8. $\frac{1}{5}\begin{bmatrix} 10 & 0 \\ 15 & -20 \end{bmatrix} = \begin{bmatrix} 2 & 0 \\ 3 & -4 \end{bmatrix}$

9. $4\begin{bmatrix} 5 \\ -1 \end{bmatrix} - 5\begin{bmatrix} 3 \\ -4 \end{bmatrix} = \begin{bmatrix} 5 \\ 16 \end{bmatrix}$

10. $2\begin{bmatrix} 1 & -2 \\ 3 & 0 \end{bmatrix} - \begin{bmatrix} -5 & 2 \\ 0 & 3 \end{bmatrix} = \begin{bmatrix} 7 & -6 \\ 6 & -3 \end{bmatrix}$

11. $A = \begin{bmatrix} 0 & -2 \\ 1 & 0 \end{bmatrix}$ $B = \begin{bmatrix} 0 & -2 \\ 1 & 0 \end{bmatrix}$. Find

(a) $A + 2B$

$\begin{bmatrix} 0 & -2 \\ 1 & 0 \end{bmatrix} + 2\begin{bmatrix} 0 & -2 \\ 1 & 0 \end{bmatrix} = \begin{bmatrix} 0 & -6 \\ 3 & 0 \end{bmatrix}$

(b) the matrix C such that $A + C = B$

$C = B - A = \begin{bmatrix} 0 & -2 \\ 1 & 0 \end{bmatrix} - \begin{bmatrix} 0 & -2 \\ 1 & 0 \end{bmatrix}$

$= \begin{bmatrix} 0 & 0 \\ 0 & 0 \end{bmatrix}$

12. Find the matrix P which is such that $3P + \begin{bmatrix} 0 & 6 \\ -9 & 2 \end{bmatrix} = 4\begin{bmatrix} 3 & 0 \\ 0 & -4 \end{bmatrix}$

$3P = 4\begin{bmatrix} 3 & 0 \\ 0 & -4 \end{bmatrix} - \begin{bmatrix} 0 & 6 \\ -9 & 2 \end{bmatrix}$

$3P = \begin{bmatrix} 12 & -6 \\ 9 & -18 \end{bmatrix}$

$P = \begin{bmatrix} 4 & -2 \\ 3 & -6 \end{bmatrix}$

6F MULTIPLICATION OF MATRICES

Matrices may be multiplied only if they are compatible; the number of columns in the first matrix equals to the number of rows in the second matrix.

To perform matrix multiplication follow the following steps:
Step 1 : Write the size of each matrix next to each other.
Step 2 : Circle the middle two numbers; if they are same multiplication is possible.
Step 3 : The remaining two numbers represent the size of the product matrix.

EXAMPLE 1

Given that $A = \begin{bmatrix} 3 & 2 \\ 1 & 0 \end{bmatrix}$ and $B = \begin{bmatrix} 5 \\ 4 \end{bmatrix}$, express AB as a single matrix.

SOLUTION

$$\begin{bmatrix} 3 & 2 \\ 1 & 0 \end{bmatrix} \times \begin{bmatrix} 5 \\ 4 \end{bmatrix}$$

$$2 \times \boxed{2 \quad 2} \times 1$$

The circled 2's means that multiplication is possible. The size of the answer must be 2×1.

$$\begin{bmatrix} 3 & 2 \\ 1 & 0 \end{bmatrix} \times \begin{bmatrix} 5 \\ 4 \end{bmatrix} = \begin{bmatrix} 3 \times 5 + 2 \times 4 \\ 1 \times 5 + 0 \times 4 \end{bmatrix} = \begin{bmatrix} 23 \\ 5 \end{bmatrix}$$

EXAMPLE 3

Given that $A = \begin{bmatrix} 2 & 1 \\ 4 & 5 \end{bmatrix}$, simplify A^2.

SOLUTION

$A^2 = A \times A = \begin{bmatrix} 2 & 1 \\ 4 & 5 \end{bmatrix} \times \begin{bmatrix} 2 & 1 \\ 4 & 5 \end{bmatrix}$

$= \begin{bmatrix} 2 \times 2 + 1 \times 4 & 2 \times 1 + 1 \times 5 \\ 4 \times 2 + 5 \times 4 & 4 \times 1 + 5 \times 5 \end{bmatrix}$

$= \begin{bmatrix} 8 & 7 \\ 28 & 29 \end{bmatrix}$

EXAMPLE 2

Given that $A = \begin{bmatrix} 2 & 1 \\ 4 & 3 \end{bmatrix}$ and $B = \begin{bmatrix} 0 & 5 \\ 8 & 7 \end{bmatrix}$, express AB as a single matrix.

SOLUTION

$$\begin{bmatrix} 2 & 1 \\ 4 & 3 \end{bmatrix} \times \begin{bmatrix} 0 & 5 \\ 8 & 7 \end{bmatrix}$$

$$2 \times \boxed{2 \quad 2} \times 2$$

$$\begin{bmatrix} 2 & 1 \\ 4 & 3 \end{bmatrix} \times \begin{bmatrix} 0 & 5 \\ 8 & 7 \end{bmatrix} = \begin{bmatrix} 2 \times 0 + 1 \times 8 & 2 \times 5 + 1 \times 7 \\ 4 \times 0 + 3 \times 8 & 4 \times 5 + 3 \times 7 \end{bmatrix}$$

$$= \begin{bmatrix} 8 & 17 \\ 24 & 41 \end{bmatrix}$$

EXAMPLE 4

Given that $\begin{bmatrix} 3 & 2 & -1 \\ 1 & 0 & 4 \end{bmatrix} \begin{bmatrix} 1 \\ m \\ 3 \end{bmatrix} = \begin{bmatrix} 4 \\ 13n \end{bmatrix}$, find the values of m and n.

SOLUTION

$\begin{bmatrix} 3 & 2 & -1 \\ 1 & 0 & 4 \end{bmatrix} \begin{bmatrix} 1 \\ m \\ 3 \end{bmatrix} = \begin{bmatrix} 3 \times 1 + 2 \times m - 1 \times 3 \\ 1 \times 1 + 0 \times m + 4 \times 3 \end{bmatrix}$

$= \begin{bmatrix} 3 + 2m - 3 \\ 1 + 0m + 12 \end{bmatrix}$

$= \begin{bmatrix} 2m \\ 13 \end{bmatrix}$

$\begin{bmatrix} 2m \\ 13 \end{bmatrix} = \begin{bmatrix} 4 \\ 13n \end{bmatrix}$

$\therefore m = 2 \text{ and } n = 1$

EXERCISE 6F

Evaluate the following matrix products

1. $\begin{bmatrix} 5 & 4 \\ 1 & 2 \end{bmatrix} \begin{bmatrix} 3 \\ 2 \end{bmatrix} = \begin{bmatrix} 23 \\ 7 \end{bmatrix}$

2. $\begin{bmatrix} 6 & 7 \\ 1 & 3 \end{bmatrix} \begin{bmatrix} 1 \\ 5 \end{bmatrix} = \begin{bmatrix} 41 \\ 16 \end{bmatrix}$

3. $\begin{bmatrix} 5 & 4 \\ 1 & 0 \end{bmatrix} \begin{bmatrix} 3 & 6 \\ 4 & 3 \end{bmatrix} = \begin{bmatrix} 31 & 42 \\ 3 & 6 \end{bmatrix}$

4. $\begin{bmatrix} 7 & 2 \\ 4 & 0 \end{bmatrix} \begin{bmatrix} 5 & -2 \\ 1 & 3 \end{bmatrix} = \begin{bmatrix} 37 & -8 \\ 20 & -8 \end{bmatrix}$

5. $\begin{bmatrix} 4 & 5 \end{bmatrix} \begin{bmatrix} -3 \\ 2 \end{bmatrix} = \begin{bmatrix} -2 \end{bmatrix}$

6. $\begin{bmatrix} 4 & 0 & 5 \end{bmatrix} \begin{bmatrix} -3 \\ 2 \\ 1 \end{bmatrix} = \begin{bmatrix} -7 \end{bmatrix}$

7. $\begin{bmatrix} 3 \\ 6 \end{bmatrix} \begin{bmatrix} 2 & 5 \end{bmatrix} = \begin{bmatrix} 6 & 15 \\ 12 & 30 \end{bmatrix}$

8. $\begin{bmatrix} 4 \\ -2 \end{bmatrix} \begin{bmatrix} 7 & 10 \end{bmatrix} = \begin{bmatrix} 28 & 40 \\ -14 & -20 \end{bmatrix}$

9. $\begin{bmatrix} 8 & -1 & 1 \\ 4 & 0 & 2 \end{bmatrix} \begin{bmatrix} 2 \\ 5 \\ 3 \end{bmatrix} = \begin{bmatrix} 14 \\ 14 \end{bmatrix}$

10. $\begin{bmatrix} 3 & 4 \end{bmatrix} \begin{bmatrix} 5 & 0 \\ -1 & 2 \end{bmatrix} = \begin{bmatrix} 11 & 8 \end{bmatrix}$

11. $\begin{bmatrix} 4 & -2 & 5 \end{bmatrix} \begin{bmatrix} 2 & 4 \\ 5 & 0 \\ 3 & -1 \end{bmatrix} = \begin{bmatrix} 13 & 11 \end{bmatrix}$

12. $\begin{bmatrix} 2 & 0 & 3 \end{bmatrix} \begin{bmatrix} 4 & 6 \\ 1 & 3 \\ 5 & 0 \end{bmatrix} = \begin{bmatrix} 23 & 12 \end{bmatrix}$

13. $\begin{bmatrix} 5 & 4 \\ 1 & 0 \end{bmatrix} \begin{bmatrix} 1 & 0 \\ 0 & 1 \end{bmatrix} = \begin{bmatrix} 5 & 4 \\ 1 & 0 \end{bmatrix}$

14. $\begin{bmatrix} 6 & 5 \\ 4 & 3 \end{bmatrix} \begin{bmatrix} 1 & 0 \\ 0 & 1 \end{bmatrix} = \begin{bmatrix} 6 & 5 \\ 4 & 3 \end{bmatrix}$

15. Given that $\begin{bmatrix} 5 & 2 & -1 \\ 1 & 0 & 4 \end{bmatrix} \begin{bmatrix} 1 \\ m \\ 5 \end{bmatrix} = \begin{bmatrix} 8 \\ 7n \end{bmatrix}$, find the values of m and n.

$$\begin{bmatrix} 5 + 2m - 5 \\ 1 + 20 \end{bmatrix} = \begin{bmatrix} 8 \\ 7n \end{bmatrix}$$

$2m = 8$
$m = 4$
$7n = 21$
$n = 3$

16. Given that $\begin{bmatrix} 3 & 1 \\ -1 & q \end{bmatrix} \begin{bmatrix} 1 & 2 \\ 0 & 1 \end{bmatrix} = \begin{bmatrix} p & 7 \\ -1 & 4 \end{bmatrix}$, find the value of p and the value of q.

$$\begin{bmatrix} 3 & 7 \\ -1 & -2+q \end{bmatrix} = \begin{bmatrix} p & 7 \\ -1 & 4 \end{bmatrix}$$

$p = 3$
$-2 + q = 4$
$q = 6$

17. $M = \begin{bmatrix} 1 & s \\ r & 6 \end{bmatrix}$ and $N = \begin{bmatrix} 2 & -3 \\ 0 & 8 \end{bmatrix}$

(a) Express 4M − 3N in terms of r and s.

$4 \begin{bmatrix} 1 & s \\ r & 6 \end{bmatrix} - 3 \begin{bmatrix} 2 & -3 \\ 0 & 8 \end{bmatrix}$

$= \begin{bmatrix} 4 & 4s \\ 4r & 24 \end{bmatrix} - \begin{bmatrix} 6 & -9 \\ 0 & 24 \end{bmatrix}$

$= \begin{bmatrix} -2 & 4s+9 \\ 4r & 0 \end{bmatrix}$

(b) Find N^2.

$\begin{bmatrix} 2 & -3 \\ 0 & 8 \end{bmatrix} \times \begin{bmatrix} 2 & -3 \\ 0 & 8 \end{bmatrix} = \begin{bmatrix} 4 & -30 \\ 0 & 64 \end{bmatrix}$

(c) Given that NM = 8M, find the value of r and the value of s.

$\begin{bmatrix} 2 & -3 \\ 0 & 8 \end{bmatrix} \times \begin{bmatrix} 1 & s \\ r & 6 \end{bmatrix} = 8 \begin{bmatrix} 1 & s \\ r & 6 \end{bmatrix}$

$\begin{bmatrix} 2-3r & 2s-18 \\ 8r & 48 \end{bmatrix} = \begin{bmatrix} 8 & 8s \\ 8r & 48 \end{bmatrix}$

$2 - 3r = 8 \therefore r = -2$

$2s - 18 = 8s \therefore s = -3$

6G DETERMINING THE POWER OF A MATRIX USING TECHNOLOGY

In this part of the chapter, we are going to make use of technology (CAS) to evaluate the power of matrices where complicated numerical manipulation is required.

EXAMPLES

1. Given that $M = \begin{bmatrix} 1 & 2 & 0 \\ 0 & 5 & 4 \\ -1 & 3 & 6 \end{bmatrix}$, using your CAS to simplify M^2.

SOLUTION

$$M^2 = \begin{bmatrix} 1 & 12 & 8 \\ -4 & 37 & 44 \\ -7 & 31 & 48 \end{bmatrix}$$

2. Given that $N = \begin{bmatrix} 1 & 4 & 3 \\ 2 & 0 & 4 \\ -1 & 5 & -3 \end{bmatrix}$, using your CAS to simplify N^3.

SOLUTION

$$N^3 = \begin{bmatrix} 60 & 98 & 82 \\ 56 & -6 & 148 \\ 18 & 192 & -116 \end{bmatrix}$$

3. Given that $A = \begin{bmatrix} 3 & 1 & 0 \\ 0 & 6 & 4 \\ -1 & 4 & 3 \end{bmatrix}$ and $B = \begin{bmatrix} 2 & 4 & 3 \\ 5 & 0 & 1 \\ 1 & 8 & -2 \end{bmatrix}$, simplify $A^2 + B^2$.

SOLUTION

$$A^2 + B^2 = \begin{bmatrix} 36 & 41 & 8 \\ 7 & 80 & 49 \\ 34 & 23 & 40 \end{bmatrix}$$

EXERCISE 6G

1. Given that $P = \begin{bmatrix} 2 & 3 & 5 \\ 0 & 4 & 1 \\ 1 & 3 & 0 \end{bmatrix}$, simplify P^2.

$\begin{bmatrix} 9 & 33 & 13 \\ 1 & 19 & 4 \\ 2 & 15 & 8 \end{bmatrix}$

2. Simplify A^2 if $A = \begin{bmatrix} 1 & 8 & 9 & 2 \\ 1 & 7 & 5 & 1 \\ 2 & 0 & 5 & 0 \\ 3 & 4 & 1 & 7 \end{bmatrix}$

$\begin{bmatrix} 33 & 72 & 96 & 24 \\ 21 & 61 & 70 & 16 \\ 12 & 16 & 43 & 4 \\ 30 & 80 & 59 & 59 \end{bmatrix}$

3. Given that $Q = \begin{bmatrix} 2 & -1 & 6 & 3 \\ 1 & 3 & 1 & 1 \\ 4 & 1 & 4 & 2 \\ 3 & 2 & 3 & 4 \end{bmatrix}$, simplify Q^2.

$\begin{bmatrix} 36 & 7 & 44 & 29 \\ 12 & 11 & 16 & 12 \\ 31 & 7 & 47 & 29 \\ 32 & 14 & 44 & 33 \end{bmatrix}$

4. Simplify $A^2 + B^2$, where $A = \begin{bmatrix} 4 & 2 & 1 \\ 1 & 3 & 4 \\ 5 & 1 & 0 \end{bmatrix}$ and $B = \begin{bmatrix} 0 & 2 & 1 \\ 1 & 0 & 0 \\ 1 & 1 & 0 \end{bmatrix}$

$\begin{bmatrix} 26 & 16 & 12 \\ 27 & 17 & 14 \\ 22 & 15 & 10 \end{bmatrix}$

EXERCISE 6H

Write down the adjacency matrix for each of the following network.

	Network	Adjacency matrix
1.	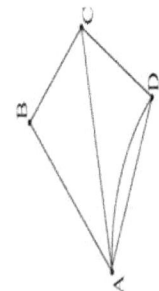	$\begin{bmatrix} 0 & 2 & 1 & 2 \\ 2 & 0 & 1 & 0 \\ 1 & 1 & 0 & 1 \\ 2 & 0 & 1 & 0 \end{bmatrix}$
2.		$\begin{bmatrix} 0 & 3 & 0 & 1 \\ 3 & 0 & 1 & 1 \\ 0 & 1 & 0 & 1 \\ 1 & 1 & 1 & 0 \end{bmatrix}$
3.		$\begin{bmatrix} 0 & 1 & 0 & 1 \\ 1 & 0 & 1 & 1 \\ 0 & 1 & 0 & 3 \\ 1 & 1 & 3 & 0 \end{bmatrix}$
4.		$\begin{bmatrix} 0 & 1 & 0 & 1 \\ 1 & 0 & 1 & 2 \\ 0 & 1 & 0 & 0 \\ 1 & 1 & 1 & 0 \end{bmatrix}$
5.		$\begin{bmatrix} 0 & 1 & 1 \\ 1 & 0 & 2 \\ 1 & 1 & 0 \end{bmatrix}$

CHAPTER 6 : MATRICES SOLUTIONS

6I TWO STAGE ADJACENCY MATRIX

A two stage adjacency matrix shows the number of two-length path from one node to the others.

Consider the network on the right showing the relationship between 3 friends A, B and C.
A considers B to be a friend but B doesn't feel the same. Likewise, B considers C to be a friend while C feels otherwise.

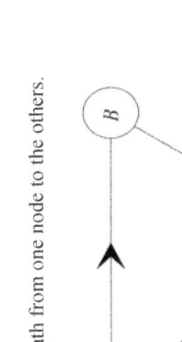

An adjacency matrix for the given network is

		FROM		
		A	B	C
TO	A	0	1	0
	B	0	0	1
	C	1	0	0

Consider a two stage matrix.

There is 1 path from A to C ($A \to B \to C$), 0 from A to B and 0 from A to itself.
Similarly, there is 1 path from B to A ($B \to C \to A$), 0 from B to C and 0 from B to itself.
And there is 1 path from C to B ($C \to A \to B$), 0 from C to A and 0 from C to itself.
Hence a two stage adjacency matrix is given as under:

		FROM		
		A	B	C
TO	A	0	0	1
	B	1	0	0
	C	0	1	0

Let $M = \begin{bmatrix} 0 & 1 & 0 \\ 0 & 0 & 1 \\ 1 & 0 & 0 \end{bmatrix}$ be the adjacency matrix.

Now $M^2 = \begin{bmatrix} 0 & 0 & 1 \\ 1 & 0 & 0 \\ 0 & 1 & 0 \end{bmatrix}$ and this is equivalent the two stage adjacency matrix.

In the examples that follow, we are going to investigate if squaring the one stage adjacency matrix is equivalent to the two stage adjacency matrix.

EXAMPLE

For the network,

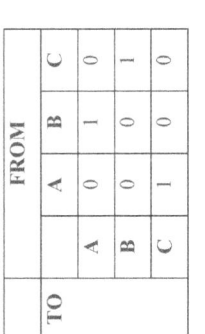

(a) Write down the adjacency matrix M.
(b) Write down the two stage adjacency matrix N.
(c) Calculate the square of the adjacency matrix M^2.
(d) Compare your answers to (b) and (c).

SOLUTION

(a) Adjacency matrix M is

		FROM			
		A	B	C	D
TO	A	0	1	0	1
	B	1	0	1	1
	C	0	1	0	1
	D	1	1	1	0

Or $M = \begin{bmatrix} 0 & 1 & 0 & 1 \\ 1 & 0 & 1 & 1 \\ 0 & 1 & 0 & 1 \\ 1 & 1 & 1 & 0 \end{bmatrix}$

(b) Two stage adjacency matrix is

		FROM			
		A	B	C	D
TO	A	2	1	2	1
	B	1	3	1	2
	C	2	1	2	1
	D	1	2	1	3

> There are 2 paths from A to itself ($A \to B \to A$ & $A \to D \to A$), 1 path from A to B ($A \to D \to B$), 2 paths from A to C ($A \to B \to C$ & $A \to D \to C$) and 1 path from A to D ($A \to B \to D$).

Or $N = \begin{bmatrix} 2 & 1 & 2 & 1 \\ 1 & 3 & 1 & 2 \\ 2 & 1 & 2 & 1 \\ 1 & 2 & 1 & 3 \end{bmatrix}$

(c) $M^2 = \begin{bmatrix} 2 & 1 & 2 & 1 \\ 1 & 3 & 1 & 2 \\ 2 & 1 & 2 & 1 \\ 1 & 2 & 1 & 3 \end{bmatrix}$

(d) Clearly $M^2 = N$, implying that a two stage matrix is equivalent to the square of a one stage adjacency matrix.

CHAPTER 6 : MATRICES SOLUTIONS

EXERCISE 6I

For each of the following determine the one stage adjacency matrix and the two stage adjacency matrix. Investigate using your CAS calculator whether the square of the one stage adjacency matrix is equivalent to the two stage adjacency matrix.

	One stage adjacency matrix	Two stage adjacency matrix
1.	$\begin{bmatrix} 0 & 1 & 1 \\ 0 & 0 & 1 \\ 0 & 0 & 0 \end{bmatrix}$	$\begin{bmatrix} 0 & 0 & 1 \\ 0 & 0 & 0 \\ 0 & 0 & 0 \end{bmatrix}$
2.	$\begin{bmatrix} 0 & 1 & 1 & 1 \\ 0 & 0 & 1 & 0 \\ 1 & 1 & 0 & 1 \\ 1 & 0 & 1 & 0 \end{bmatrix}$	$\begin{bmatrix} 2 & 1 & 2 & 1 \\ 1 & 1 & 0 & 1 \\ 1 & 1 & 3 & 1 \\ 1 & 2 & 1 & 2 \end{bmatrix}$
3.	$\begin{bmatrix} 0 & 1 & 0 & 1 \\ 0 & 0 & 1 & 0 \\ 0 & 0 & 0 & 0 \\ 1 & 0 & 1 & 0 \end{bmatrix}$	$\begin{bmatrix} 1 & 0 & 2 & 0 \\ 0 & 0 & 1 & 0 \\ 0 & 0 & 0 & 0 \\ 0 & 2 & 0 & 1 \end{bmatrix}$
4.	$\begin{bmatrix} 0 & 1 & 1 & 1 \\ 0 & 0 & 0 & 0 \\ 0 & 0 & 0 & 0 \\ 0 & 0 & 1 & 0 \end{bmatrix}$	$\begin{bmatrix} 0 & 0 & 2 & 0 \\ 0 & 0 & 0 & 0 \\ 0 & 0 & 0 & 0 \\ 0 & 0 & 0 & 0 \end{bmatrix}$

6J SOCIAL INTERACTION AS A MATRIX

Matrix as we have seen so far has various uses and importance. In this section, we are going to show that matrix can be used to portray social interaction within a social group.

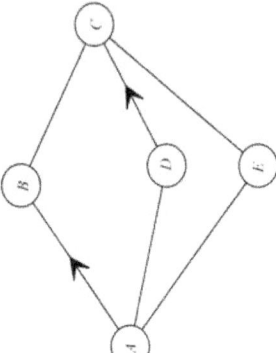

The diagram on the right shows a group of four students meeting at an orientation day at UWA. All the four students have enrolled in the same course. Student A happens to have met all the other 3 student's parents. Student B, on the other hand, knew only student C's parents prior to this meeting. Student C had already met the parents of student A, B and D.

Let us assume the number 0 indicates someone did not meet the other's parent before and the number 1 indicating they have already met their parents. The information can then be shown as a matrix as under:

		FROM			
		A	B	C	D
TO	A	0	1	1	1
	B	0	0	1	0
	C	1	1	0	1
	D	1	0	1	0

OR $S = \begin{bmatrix} 0 & 1 & 1 & 1 \\ 0 & 0 & 1 & 0 \\ 1 & 1 & 0 & 1 \\ 1 & 0 & 1 & 0 \end{bmatrix}$

It is quite obvious that someone should have met his or her own parent at one point. This is the reason we have zeroes along the leading diagonal.

EXERCISE 6J

1. Five middle school students did a survey asking in discretion "Do you like me?". The result is shown in the diagram on the right. Use 1 to represent someone liking the other and 0 for disliking.
Complete the matrix below.

	A	B	C	D	E
A	0	1	0	1	1
B	0	0	1	0	0
C	0	1	0	0	1
D	1	0	1	0	0
E	1	0	1	0	0

CHAPTER 6 : MATRICES SOLUTIONS

2. A year 12 IT teacher did a survey from his top 6 students to find out who emailed each other before. An analysis of the responses produced the diagram as shown.

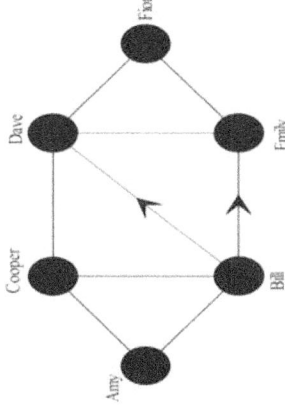

Complete the matrix below using

- 0 to show someone has emailed the other.
- 1 to show someone hasn't emailed the other
- 0 for someone emailing himself or herself.

	A	B	C	D	E	F
A	0	1	0	0	1	1
B	1	0	1	1	0	0
C	0	1	0	1	1	0
D	0	1	1	0	1	1
E	0	0	1	0	0	1
F	1	0	0	1	1	0

3. The diagram on the right shows "who has been on camp with whom in a particular year level" among a group of five classmates Alex, Bob, Carl, David and Ethan.
Complete the matrix below using

- 0 to show a pair has not been on camp together,
- 1 to show a pair has been on camp together,
- 0 for someone being on camp accompanying himself or herself.

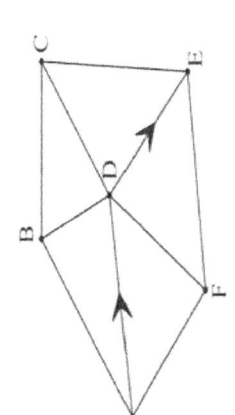

	Alex	Bob	Carl	David	Ethan
Alex	0	1	0	1	0
Bob	1	0	1	1	1
Carl	0	1	0	0	1
David	1	1	0	0	1
Ethan	0	1	1	1	0

MATHEMATICS APPLICATIONS UNIT 1

4. "Do you have my mobile number?"
The diagram shows the result of the above question asked to six friends who all work together at a fast food place.

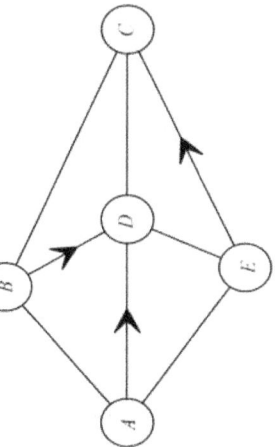

Complete the matrix below using
- 0 showing not having mobile number,
- 1 showing having mobile number,
- 0 for having someone's own mobile number.

	Amy	Bill	Cooper	Dave	Emily	Fiona
Amy	0	1	1	0	0	0
Bill	1	0	1	1	1	0
Cooper	1	1	0	1	0	0
Dave	0	0	1	0	1	1
Emily	0	0	0	1	0	1
Fiona	0	0	0	1	0	0

5. "Have you slept over at my house?" Five friends' response to the above question is shown in the diagram on the right.
Complete the matrix below using
- 0 showing not slept over,
- 1 showing having slept over,
- 0 for having slept at own house.

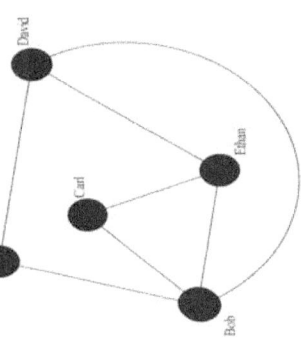

	A	B	C	D	E
A	0	1	0	1	1
B	1	0	1	1	0
C	0	1	0	1	0
D	0	0	1	0	1
E	1	0	1	1	0

6K STORING, DISPLAYING INFORMATION, MODEL AND SOLVE PROBLEMS

EXAMPLE 1

The table shows the number of games played and the results of four teams in a football league.

Team	Played	Won	Drawn	Lost
City	8	4	1	3
United	7	3	0	4
Tigers	8	4	0	4
Wolf	7	2	1	4

A win earns 3 points, a draw 1 point and a loss 0 point. Write down two matrices which on multiplication display in their product the total number of points earned by each team and hence calculate these totals.

SOLUTION

$$\begin{bmatrix} 4 & 1 & 3 \\ 3 & 0 & 4 \\ 4 & 0 & 4 \\ 2 & 1 & 4 \end{bmatrix} \begin{bmatrix} 3 \\ 1 \\ 0 \end{bmatrix} = \begin{bmatrix} 13 \\ 9 \\ 12 \\ 7 \end{bmatrix} \quad \text{or} \quad \begin{bmatrix} 3 & 1 & 0 \end{bmatrix} \begin{bmatrix} 4 & 3 & 4 & 2 \\ 1 & 0 & 0 & 1 \\ 3 & 4 & 4 & 4 \end{bmatrix} = \begin{bmatrix} 13 & 9 & 12 & 7 \end{bmatrix}$$

EXAMPLE 2

A circus show is held over a three-day period – Friday, Saturday and Sunday. The table below shows the entry price per day for an adult and for a child, and the number of adults and children attending on each day.

	Friday	Saturday	Sunday
Price ($) - Adult	12	10	10
Price ($) – Child	6	5	5
Number of adults	360	190	400
Number of children	50	60	150

(i) Write down two matrices such that their product will give the amount of entry money paid on Friday and hence calculate this product.

$$\begin{bmatrix} 360 & 50 \end{bmatrix} \begin{bmatrix} 12 \\ 6 \end{bmatrix} = \begin{bmatrix} 4620 \end{bmatrix} \quad \text{or} \quad \begin{bmatrix} 12 & 6 \end{bmatrix} \begin{bmatrix} 360 \\ 50 \end{bmatrix} = \begin{bmatrix} 4620 \end{bmatrix}$$

(ii) Write down two matrices such that their product will give the amount of entry money paid on Sunday and hence calculate this product.

$$\begin{bmatrix} 400 & 150 \end{bmatrix} \begin{bmatrix} 10 \\ 5 \end{bmatrix} = \begin{bmatrix} 4750 \end{bmatrix}$$

(iii) Calculate the percentage increase in revenue on Sunday compared to Friday.

$$\frac{4750 - 4620}{4620} \times 100 = 2.81\%$$

EXERCISE 6K

1. The table shows the results achieved by three teams in twelve events of an athletics competition. In each event, 1st place scores 5 points, 2nd place scores 3 points, and 3rd place scores 1 point.

Place / Team	1st	2nd	3rd
Hobbits	4	5	3
Dwarfs	6	0	6
Hogwarts	2	7	3

(i) Write down two matrices whose product shows the total number of points scored by each team.

$$\begin{bmatrix} 4 & 5 & 3 \\ 6 & 0 & 6 \\ 2 & 7 & 3 \end{bmatrix} \begin{bmatrix} 5 \\ 3 \\ 1 \end{bmatrix}$$

(ii) Evaluate this product of matrices.

$$\begin{bmatrix} 4 \times 5 + 5 \times 3 + 3 \times 1 \\ 6 \times 5 + 0 \times 3 + 6 \times 1 \\ 2 \times 5 + 7 \times 3 + 3 \times 1 \end{bmatrix} = \begin{bmatrix} 38 \\ 36 \\ 34 \end{bmatrix}$$

2. During a numeracy test, students take three multiple-choice tests, each with ten questions. A correct answer earns 5 marks. If no answer is given 1 mark is scored. An incorrect answer loses 2 marks. A student's final total mark is the sum of 20% of the mark in test 1, 30% of the mark in test 2 and 50% of the mark in test 3. One student's responses are summarized in the table below.

	Test 1	Test 2	Test 3
Correct answer	7	6	5
No answer	2	0	1
Incorrect answer	1	4	4

Write down three matrices such that matrix multiplication will give this student's final total mark and hence find this total mark.

$$\begin{bmatrix} 0.2 & 0.3 & 0.5 \end{bmatrix} \begin{bmatrix} 7 & 2 & 1 \\ 6 & 0 & 4 \\ 5 & 1 & 4 \end{bmatrix} \times \begin{bmatrix} 5 \\ 1 \\ -2 \end{bmatrix} = \begin{bmatrix} 22.6 \end{bmatrix}$$

CHAPTER 6 : MATRICES SOLUTIONS

3. Western Air has a fleet of aircraft consisting of 3 aircraft of type P, 5 of type Q, 2 of type R and 10 of type S. The aircraft have 3 classes of seat known as Economy, Business and First. The table below shows the number of these seats in each of the 4 types of aircraft.

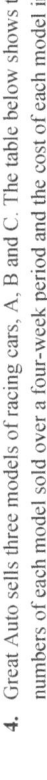

	Economy	Business	First
P	300	30	10
Q	180	40	15
R	140	20	5
S	110	5	0

(i) Write down two matrices whose product shows the total number of seats in each class.

$$\begin{bmatrix} 300 & 180 & 140 & 110 \\ 30 & 40 & 20 & 5 \\ 10 & 15 & 5 & 0 \end{bmatrix} \times \begin{bmatrix} 3 \\ 5 \\ 2 \\ 10 \end{bmatrix}$$

(ii) Evaluate this product of matrices.

$$\begin{bmatrix} 3180 \\ 380 \\ 115 \end{bmatrix}$$

On a particular day, each aircraft made one flight. 5% of the Economy seats were empty, 10% of the Business seats were empty and 20% of the First seats were empty.

(iii) Write down a matrix whose product with the matrix found in part (ii) will give the total number of empty seats on that day.

$$[0.05 \quad 0.1 \quad 0.2] \times \begin{bmatrix} 3180 \\ 380 \\ 115 \end{bmatrix}$$

(iv) Evaluate this total.

$$[0.05 \times 3180 + 0.1 \times 380 + 0.2 \times 115] = [220]$$

MATHEMATICS APPLICATIONS UNIT 1

4. Great Auto sells three models of racing cars, A, B and C. The table below shows the numbers of each model sold over a four-week period and the cost of each model in $.

Week \ Model	A	B	C
1	8	12	4
2	7	10	2
3	10	12	0
4	6	8	4
Costs($)	250	400	700

In the first two weeks the shop banked 30% of all money received, but in the last two weeks the shop only banked 20% of all money received.

(i) Write down three matrices such that matrix multiplication will give the total amount of money banked over the four-week period.

$$[0.3 \quad 0.3 \quad 0.2 \quad 0.2] \times \begin{bmatrix} 8 & 12 & 4 \\ 7 & 10 & 2 \\ 10 & 12 & 0 \\ 6 & 8 & 4 \end{bmatrix} \times \begin{bmatrix} 250 \\ 400 \\ 700 \end{bmatrix}$$

(ii) Hence evaluate this total amount.

$$[7985]$$

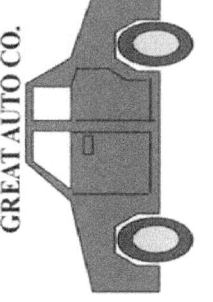

CHAPTER 7

THE THEOREM OF PYTHAGORAS

A triangle that contains a 90° angle is called a right-angled triangle as shown.

The longest side of the triangle is called the **hypotenuse**.

The Pythagoras theorem states

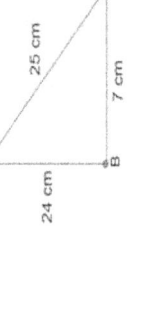

> ***For any right-angled triangle, the square of the length of the hypotenuse is equal to the sum of the squares of the lengths of the two shorter sides.***

For the triangle on the right

$AB^2 = AC^2 + BC^2$ or $c^2 = a^2 + b^2$

7A   HOW TO PROVE A TRIANGLE IS RIGHT ANGLED?

To prove a triangle is right-angled, we have to use the converse of Pythagoras theorem. In other words, the sum of the squares of the lengths of the two shorter sides must be equal the square of the length of the hypotenuse.

EXAMPLE 1

Show that the triangle is right-angled.

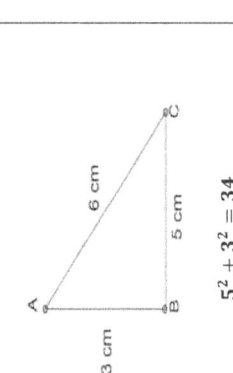

SOLUTION

Let $c = 10, a = 6$ and $b = 8$

$6^2 + 8^2 = 36 + 64 = 100$
And $10^2 = 100$.
As $c^2 = a^2 + b^2$ holds for the given triangle, it is proven that it is a right-angled triangle.

EXAMPLE 2

Show that the triangle is **NOT** right-angled.

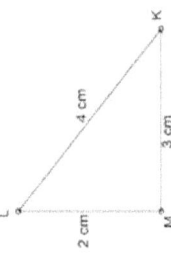

SOLUTION

Let $c = 4, a = 2$ and $b = 3$

$2^2 + 3^2 = 4 + 9 = 13$
And $4^2 = 16$.

As $c^2 = a^2 + b^2$ does not hold for the given triangle, it is proven that it is not a right-angled triangle.

EXERCISE 7A

Which of the following triangles are right-angled?

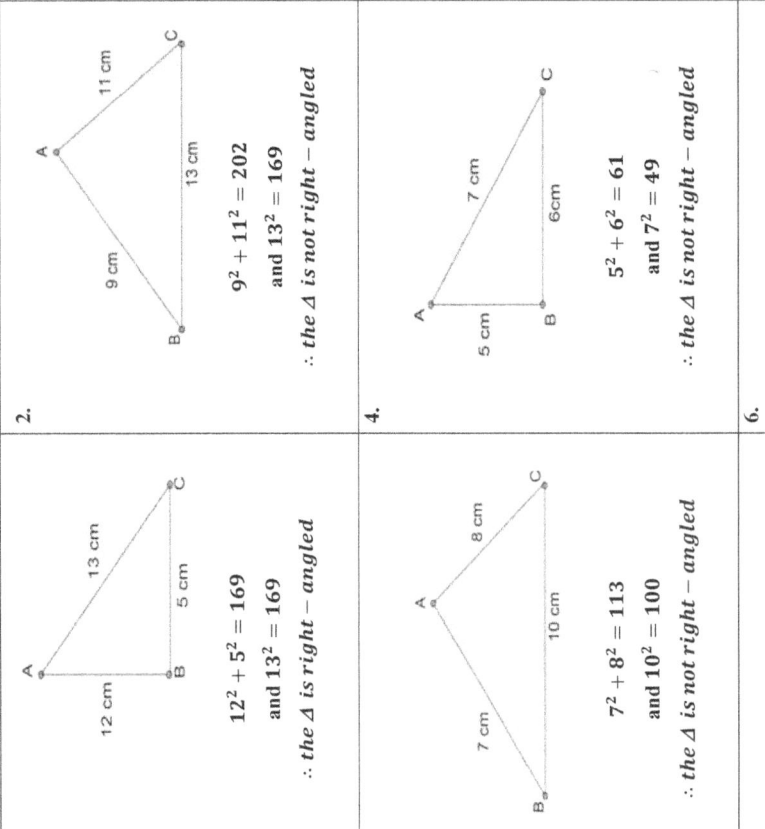

1.
$12^2 + 5^2 = 169$
and $13^2 = 169$
∴ *the Δ is right – angled*

2.
$9^2 + 11^2 = 202$
and $13^2 = 169$
∴ *the Δ is not right – angled*

3.
$7^2 + 8^2 = 113$
and $10^2 = 100$
∴ *the Δ is not right – angled*

4.
$5^2 + 6^2 = 61$
and $7^2 = 49$
∴ *the Δ is not right – angled*

5.
$5^2 + 3^2 = 34$
and $6^2 = 36$
∴ *the Δ is not right – angled*

6.
$24^2 + 7^2 = 625$
and $25^2 = 625$
∴ *the Δ is right – angled*

7B MENTAL SKILLS

CLASS ACTIVITY 1

Evaluate the following without the use of a calculator.

1. $5^2 = 25$	**2.** $7^2 = 49$	**3.** $9^2 = 81$
4. $10^2 = 100$	**5.** $30^2 = 900$	**6.** $50^2 = 2500$
7. $200^2 = 40000$	**8.** $600^2 = 360000$	**9.** $\left(\frac{4}{5}\right)^2 = \frac{16}{25}$
10. $\sqrt{9} = 3$	**11.** $\sqrt{64} = 8$	**12.** $\sqrt{121} = 11$
13. $\sqrt{169} = 13$	**14.** $\sqrt{225} = 15$	**15.** $\sqrt{4900} = 70$
16. $\sqrt{9+16} = 5$	**17.** $\sqrt{40-36} = 2$	**18.** $\sqrt{36+64} = 10$
19. $\sqrt{144+25} = 13$	**20.** $\sqrt{600-200} = 20$	**21.** $\sqrt{1200+400} = 40$

Evaluate the following to 2 decimal places, using your calculator where necessary.

22. $\sqrt{5^2 + 4^2} = 6.40$	**23.** $\sqrt{7^2 + 11^2} = 13.04$	**24.** $\sqrt{25^2 + 16^2} = 29.68$
25. $\sqrt{15^2 + 20^2} = 25$	**26.** $\sqrt{17^2 + 18^2} = 24.76$	**27.** $\sqrt{2^2 + 3^2} = 3.61$

CLASS ACTIVITY 2

For each of the following write down the relationship between the pronumerals. The first two has been done as examples.

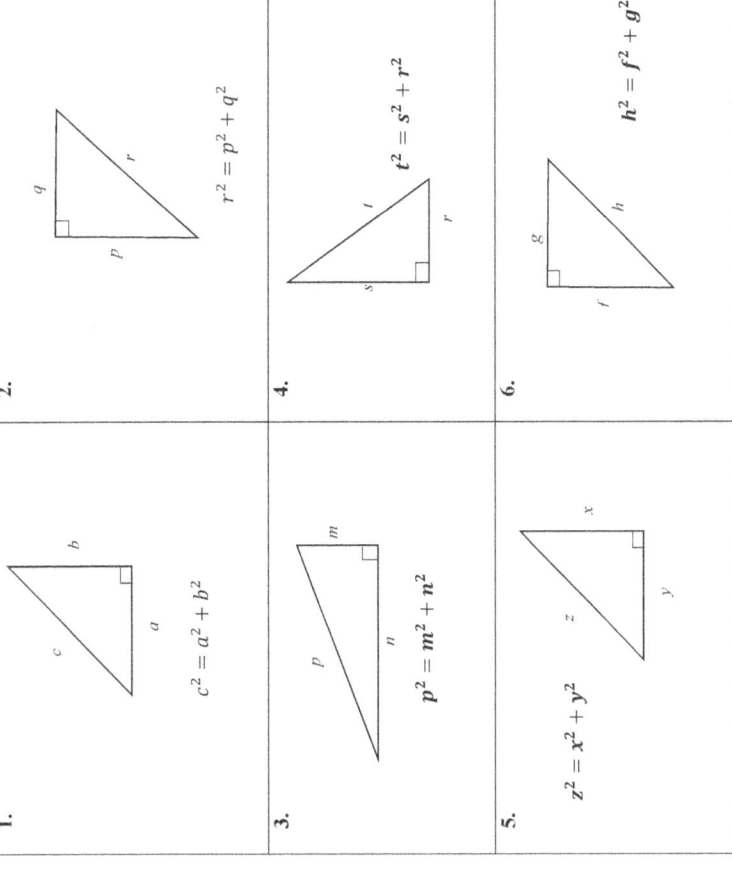

1. $c^2 = a^2 + b^2$
2. $r^2 = p^2 + q^2$
3. $p^2 = m^2 + n^2$
4. $t^2 = s^2 + r^2$
5. $z^2 = x^2 + y^2$
6. $h^2 = f^2 + g^2$

For the right-angled triangles shown below, circle the true statement.

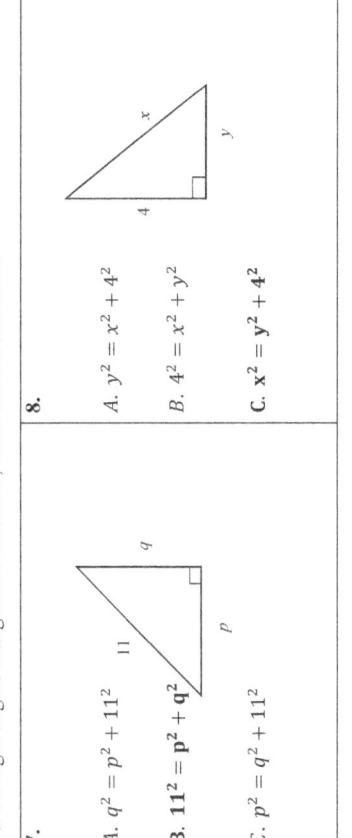

7.
A. $q^2 = p^2 + 11^2$
B. $11^2 = p^2 + q^2$
C. $p^2 = q^2 + 11^2$

8.
A. $y^2 = x^2 + 4^2$
B. $4^2 = x^2 + y^2$
C. $x^2 = y^2 + 4^2$

CHAPTER 7 : THE THEOREM OF PYTHAGORAS

7C FINDING THE LENGTH OF THE HYPOTENUSE

If two shorter sides (also known as legs) of a right-angled are known, we can use Pythagoras' Theorem to find the length of the hypotenuse.

EXAMPLE 1

Use Pythagoras' theorem to find the length of the unknown side.

SOLUTION

Using $c^2 = a^2 + b^2$

Here $a = 3, b = 4, c = x$

Substituting the values, we have

$x^2 = 3^2 + 4^2 = 9 + 16 = 25$

$\therefore x = \sqrt{25} = 5$ (ignore $x = -5$)

EXAMPLE 2

Use Pythagoras' theorem to find the length of y, giving your answer to 2 decimal places.

SOLUTION

Using $c^2 = a^2 + b^2$

Here $a = 7, b = 10, c = y$

Substituting the values, we have

$y^2 = 7^2 + 10^2 = 49 + 100 = 149$

$\therefore y = \sqrt{149} = 12.21$

EXAMPLE 3

The diagram shows a trapezium subdivided into a square of side 12 cm and a triangle. The length of the longer parallel side is 17 cm. Find the value of x.

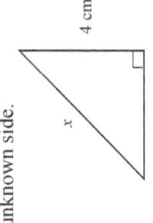

The two shorter sides of the triangle are 12 cm and $17 - 12 = 5$ cm.

SOLUTION

Using $c^2 = a^2 + b^2$

Here $a = 5, b = 12, c = x$

Substituting the values, we have

$x^2 = 5^2 + 12^2 = 25 + 144 = 169$

$\therefore x = \sqrt{169} = 13$ cm.

EXERCISE 7C

Use Pythagoras' theorem to find the length of the unknown side.

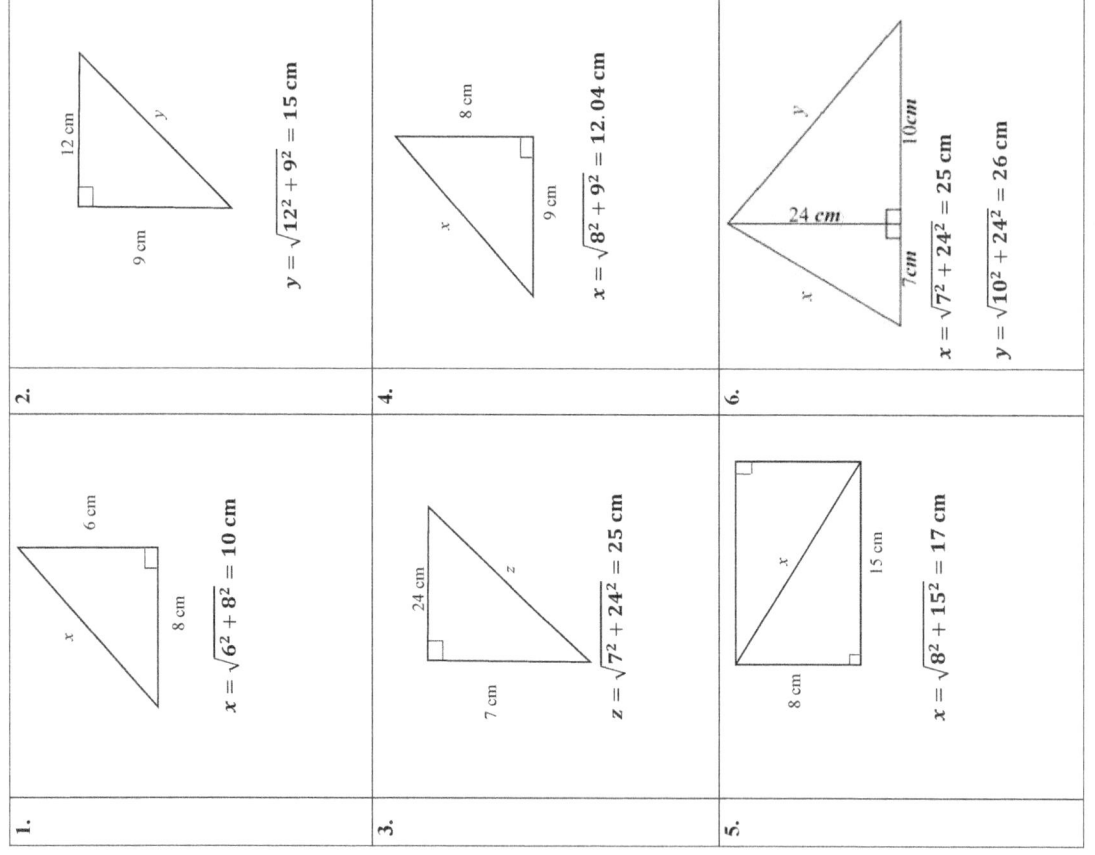

1. $x = \sqrt{6^2 + 8^2} = 10$ cm

2. $y = \sqrt{12^2 + 9^2} = 15$ cm

3. $z = \sqrt{7^2 + 24^2} = 25$ cm

4. $x = \sqrt{8^2 + 9^2} = 12.04$ cm

5. $x = \sqrt{8^2 + 15^2} = 17$ cm

6. $x = \sqrt{7^2 + 24^2} = 25$ cm
   $y = \sqrt{10^2 + 24^2} = 26$ cm

CHAPTER 7 : THE THEOREM OF PYTHAGORAS SOLUTIONS

7D FINDING THE LENGTH OF THE SHORTER SIDES

The length of one of the shorter sides of a right-angled triangle can be found if we know the length of the hypotenuse and one of the shorter sides.

EXAMPLE 1

Use Pythagoras' theorem to find the length of the unknown side.

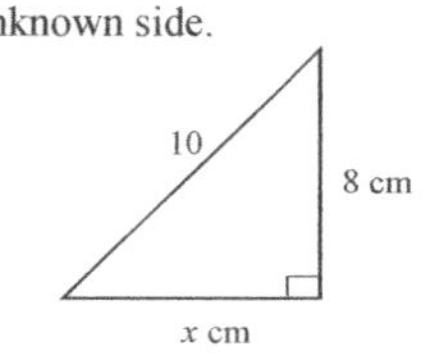

SOLUTION

Using $c^2 = a^2 + b^2$

Here $a = x, b = 8, c = 10$

Substituting the values, we have

$10^2 = x^2 + 8^2$

$x^2 = 10^2 - 8^2$

$= 100 - 64 = 36$

$\therefore x = \sqrt{36} = 6$ (*ignore* $x = -6$)

EXAMPLE 2

Use Pythagoras' theorem to find the length of x, giving your answer to 2 decimal places.

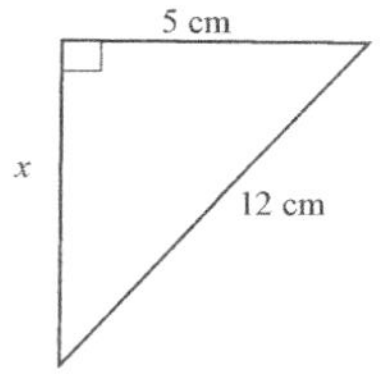

SOLUTION

Using $c^2 = a^2 + b^2$

Here $a = x, b = 5, c = 12$

Substituting the values, we have

$12^2 = x^2 + 5^2$

$x^2 = 12^2 - 5^2$

$= 144 - 25 = 119$

$\therefore x = \sqrt{119} = 10.91$

EXAMPLE 3

The diagram shows a trapezium subdivided into a square of side 10 cm and a triangle. Find the value of x.

SOLUTION

Using $c^2 = a^2 + b^2$

Here $a = x, b = 10, c = 14$

Substituting the values, we have

$14^2 = x^2 + 10^2$ $\therefore x^2 = 14^2 - 10^2$

$= 196 - 100 = 96$

$\therefore x = \sqrt{96} = 9.80 \; cm$

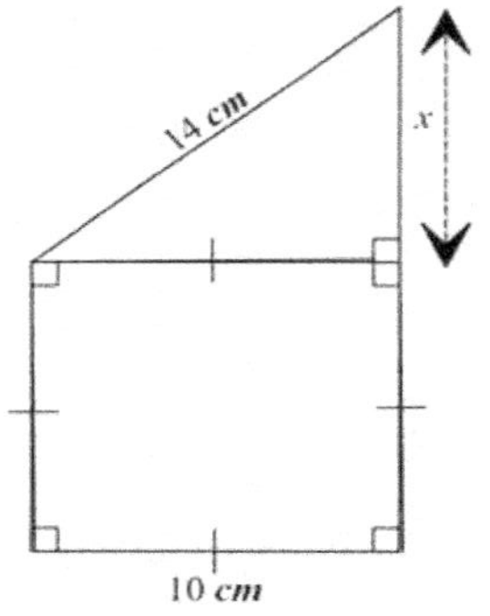

EXERCISE 7D

Use Pythagoras' theorem to find the length of the unknown side.

1.

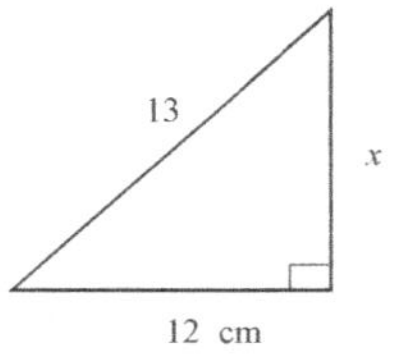

$x = \sqrt{13^2 - 12^2} = \mathbf{5 \; cm}$

2.

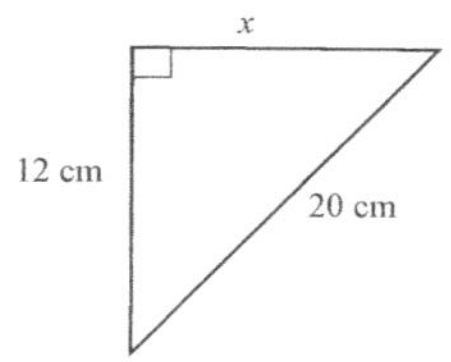

$x = \sqrt{20^2 - 12^2} = \mathbf{16 \; cm}$

3.

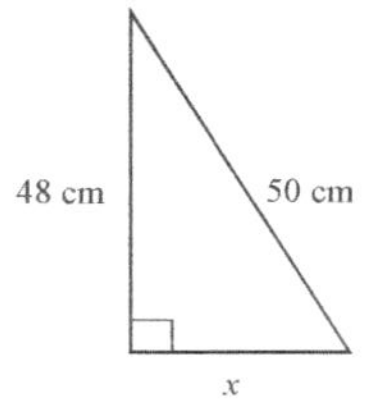

$x = \sqrt{50^2 - 48^2} = \mathbf{14 \; cm}$

4.

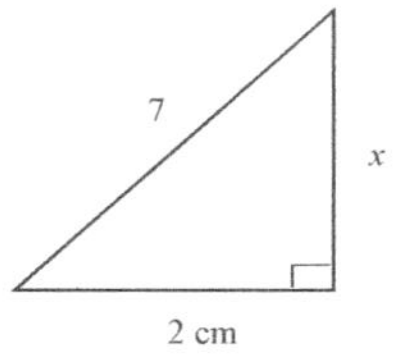

$x = \sqrt{7^2 - 2^2} = \mathbf{6.71 \; cm}$

5.

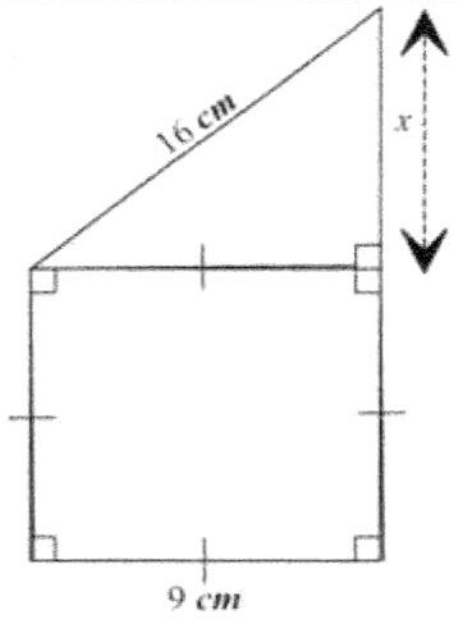

$x = \sqrt{16^2 - 9^2} = \mathbf{13.23 \; cm}$

6.

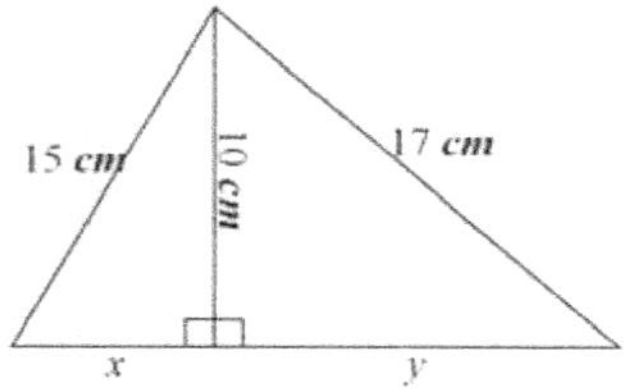

$x = \sqrt{15^2 - 10^2} = \mathbf{11.18 \; cm}$

$y = \sqrt{17^2 - 10^2} = \mathbf{13.75 \; cm}$

CHAPTER 7: THE THEOREM OF PYTHAGORAS SOLUTIONS

7E APPLICATIONS IN TWO DIMENSIONS

It is true that one can use Pythagoras theorem in maths problems. But there are other uses too. It is widely used in the world of architecture and in construction of buildings. It is also used to find the speed of sound in water. Furthermore, geologists make use of Pythagoras Theorem to determine the centre of an earthquake.

To solve an application problem with Pythagoras Theorem, follow these simple steps:

- Draw a right-angled triangle to represent the situation
- Identify the right angle, legs and the hypotenuse of the triangle
- Use variables such as x or y to label the unknown sides
- Use Pythagoras Theorem to determine the length of the unknown side.

EXAMPLE 1

A 10 m ladder leans against a wall. The foot of the ladder is 5 m from the base of a vertical wall. How high up the wall does the ladder reach?

SOLUTION

Using $c^2 = a^2 + b^2$

Here $a = h, b = 5, c = 10$

Substituting the values, we have

$10^2 = h^2 + 5^2$

$x^2 = 10^2 - 5^2 = 100 - 25 = 75$

$\therefore x = \sqrt{75} = 8.66\ m$

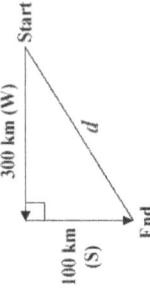

EXAMPLE 2

Sarah is practising her cycling session in regards to the upcoming Tour de France. On a Saturday morning, starting from home she cycled 300 km West and then 100 km South.

(a) Draw a well-labelled diagram to illustrate the situation.

(b) How far is she from home after the 400 km tour on bike?

Using $c^2 = a^2 + b^2$

Here $a = 3, b = 4, c = d$

$d^2 = 100^2 + 300^2 = 100\,000$

$\therefore d = \sqrt{100000} = 316.23\ km$

MATHEMATICS APPLICATIONS UNIT 1

EXERCISE 7E

1. A ladder of length 13 m is leaning against a vertical wall. The ladder reaches 12 m up the wall. How far is the foot of the ladder from the bottom of the wall on horizontal ground?

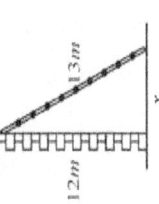

$x = \sqrt{13^2 - 12^2} = 5\ m$

2. Johnny is rapelling (descending at the end of a rope) at Bayside park. His task is to slide down a rope starting 20 m high and goes on for a horizontal distance of 50 m across the river.

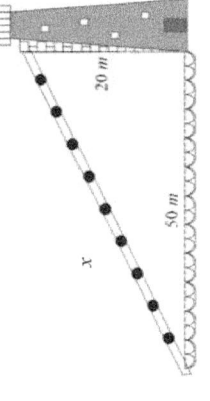

If Johnny did the rapelling 4 times, calculate the total distance he descended along the rope.

$x = \sqrt{20^2 + 50^2} = 53.85\ m$

$53.85 \times 4 = 215.4\ m$

3. Find the length of the diagonal of a rectangular swimming pool whose dimensions are 36 m by 18 m.

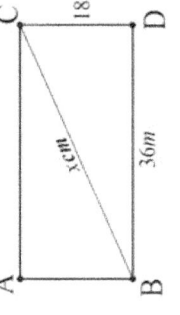

$x = \sqrt{18^2 + 36^2} = 40.25\ m$

4. As part of his training for the Formula Two competition coming soon, Alex does 50 round trips from A to B. Calculate the total distance travelled, giving your answer in km.

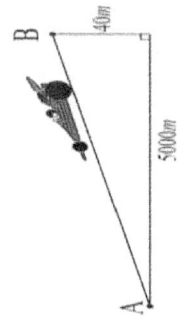

$\sqrt{5^2 + 0.04^2} \times 100 = 500.02\ km$

CHAPTER 7 : THE THEOREM OF PYTHAGORAS SOLUTIONS

5. In the diagram, D is the point on CA produced such that BD = 14.3 m. Find the length of AD.

$$BC = \sqrt{10.2^2 - 6^2} = 8.25 \text{ m}$$

$$CD = \sqrt{14.3^2 - 8.25^2} = 11.68 \text{ m}$$

$$AD = 11.68 - 6 = 5.68 \text{ m}$$

6. PQ is a chord of a circle, centre O. X is the midpoint of PQ.
OX = 6 cm and the radius of the circle is 10 cm. Calculate the length of PQ.

$$PX = \sqrt{10^2 - 6^2} = 8 \text{ cm}$$

$$PQ = 8 \times 2 = 16 \text{ cm}$$

7. Alan bought a kite for his son's 12th birthday. The measurements are shown in the diagram. The outside frame is made of Jarrah. Calculate the length of wood used.

$$x = \sqrt{24^2 + 24^2} = 33.94 \text{ cm}$$

$$y = \sqrt{24^2 + 60^2} = 64.62 \text{ cm}$$

wood used $= 33.94 \times 2 + 64.62 \times 2 = 197.12 \text{ cm}$

8. A vertical flagpole stands on a horizontal wooden base. The flagpole is held by two iron rods of lengths 16m and 22m as shown in the diagram. Use the measurements to calculate the length of LM.

$$LN = \sqrt{16^2 - 10^2} = 12.5 \text{ m}$$

$$MN = \sqrt{22^2 - 16^2} = 15.1 \text{ m}$$

$$LM = 12.5 + 15.1 = 27.6 \text{ m}$$

MATHEMATICS APPLICATIONS UNIT 1

7F APPLICATIONS IN THREE DIMENSIONS

EXAMPLES

1. The diagram shows a cylinder of height 12cm and having radius 4 cm. Find the length of the longest rod that would fit in the cylinder.

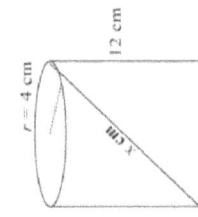

SOLUTION

The radius being 4 cm, the diameter will be 8 cm as shown.

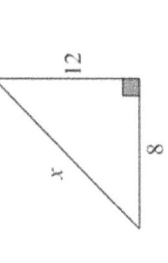

$$\therefore x = \sqrt{12^2 + 8^2} = 14.42 \text{ cm}$$

2. A right circular cone has base radius 8 m and perpendicular height 15 m. Find the slant height.

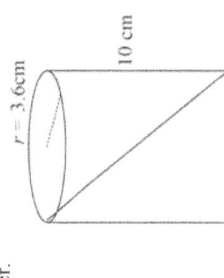

SOLUTION

$$x = \sqrt{8^2 + 15^2} = 17 \text{ m}$$

EXERCISE 7F

1. The diagram shows a rectangular prism. Work out the length of AB.

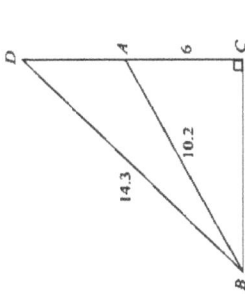

$$x = \sqrt{5^2 + 4^2} = 6.4 \text{ cm}$$

$$AB = \sqrt{9^2 + 6.4^2} = 11.04 \text{ cm}$$

2. The diagram shows a cylinder of height 10cm and having radius 3.6 cm. Find the length of the longest rod that would fit in the cylinder.

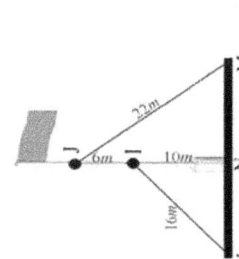

$$\sqrt{10^2 + 7.2^2} = 12.32 \text{ cm}$$

CHAPTER 7: THE THEOREM OF PYTHAGORAS SOLUTIONS

3. Use Pythagoras Theorem to calculate the length of

   (a) QR

   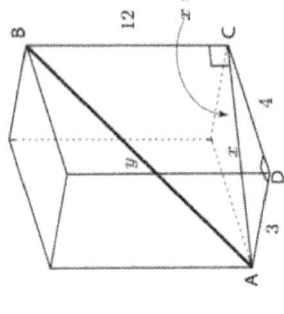

   $$QR = \sqrt{14^2 + 6^2} = 15.23 \text{ cm}$$

   (b) PQ

   $$PQ = \sqrt{15.23^2 + 4^2} = 15.75 \text{ cm}$$

4. Thomas just had a shed installed in his backyard. He is planning to lay Christmas lights along the edge AB. Calculate the length of lights thread he will need.

   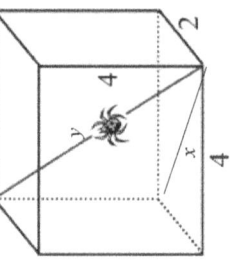

   $$AB = \sqrt{7^2 + 5.2^2} = 8.72 \text{ m}$$

5. Use Pythagoras Theorem to calculate

   (a) CE,

   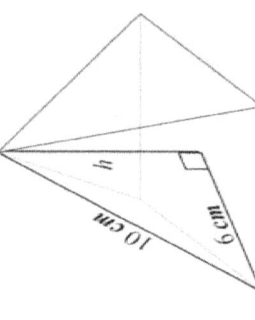

   $$CE = \sqrt{13.6^2 + 6.8^2} = 15.21 \text{ m}$$

   (b) CD.

   $$CD = \sqrt{15.21^2 + 5^2} = 16.01 \text{ m}$$

6. Peter's Christmas present box measures 4 cm long, 3 cm wide and 12 cm high. Peter was investigating the longest piece of candy that could fit in his box. He was able to figure out that the candy must start at A and end at B as shown. Use Pythagoras Theorem to find the values of x and y.

   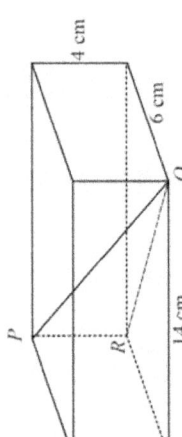

   $$x = \sqrt{3^2 + 4^2} = 5 \text{ cm}$$

   $$y = \sqrt{12^2 + 5^2} = 13 \text{ cm}$$

7. A spider crawls along a wooden stick placed inside a rectangular prism of length 4 m, width 2 m and height 4 m as shown. How far does the spider crawl if it goes from one end to the other end of the stick.

   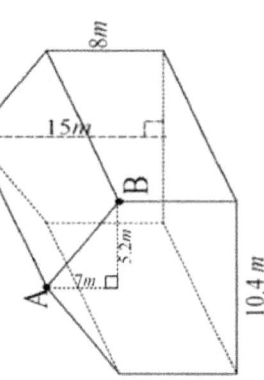

   $$x = \sqrt{2^2 + 4^2} = 4.47 \text{ m}$$

   $$y = \sqrt{4.47^2 + 4^2} = 6 \text{ m}$$

8. The base of a pyramid is a square with diagonals of length 12 cm. The sloping faces are isosceles triangles with equal sides of length 10 cm. The height of the pyramid is h cm. Calculate h.

   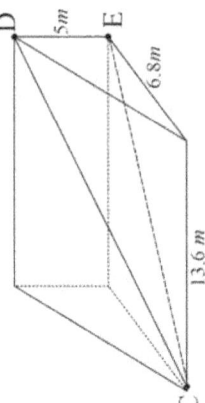

   $$h = \sqrt{10^2 - 6^2} = 8 \text{ cm}$$

CHAPTER 8

PERIMETER AND AREA

8A PERIMETER

Perimeter is the total distance around the outside of a two-dimensional (2D) shape. To find the perimeter of a shape, we need to add the length of all the sides, including curved sides in some cases. Perimeter can be measured in different units such as metres, inches, km, feet etc..

EXERCISE 8A

Calculate the perimeter of these shapes:

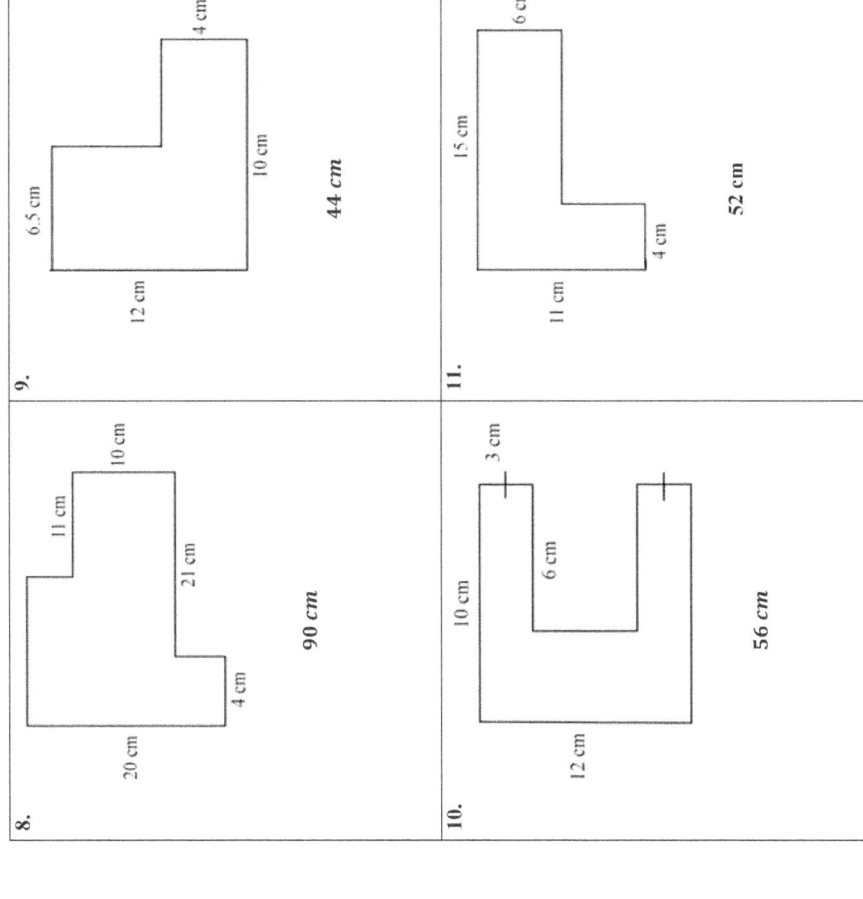

1. $10 + 13.2 + 7.8 = 31\ cm$

2. $2.6 + 4.1 + 11.5 + 4.3 + 5.4 + 2.1 = 30\ cm$

3. $6.3 \times 5 = 31.5\ cm$

4. $7.4 + 11.2 + 8.1 + 4.3 = 31\ cm$

5. $9 + 7 + 4.5 + 2.5 + 3 = 26\ cm$

6. $2(7.5 + 4.2) = 23.4\ cm$

7. The diagram shows a rectangle and a square.
   If they have equal perimeters, what is the length of one side of the square?

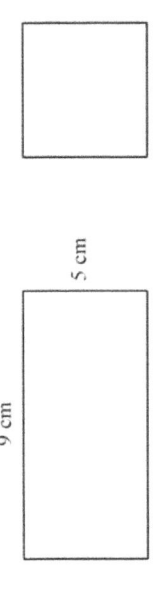

$Perimeter\ of\ rectangle = 2(9+5) = 28\ cm$

$\therefore Square\ has\ side = \dfrac{28}{4} = 7\ cm$

In these composite shapes all the lengths are in centimetres. Work out the perimeter.

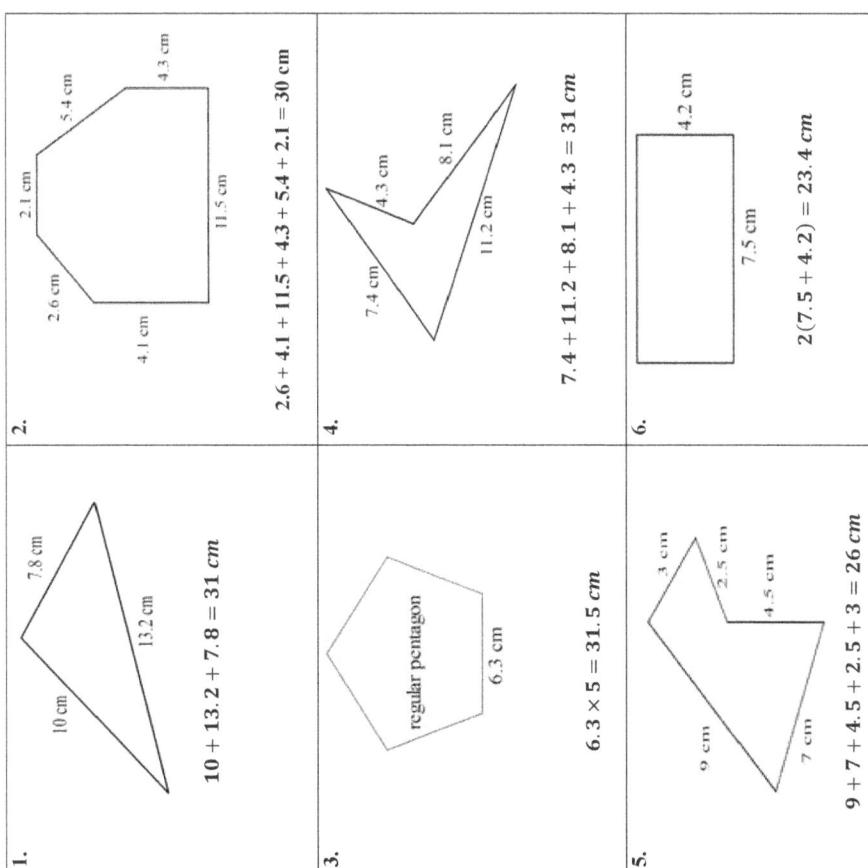

8. $90\ cm$

9. $44\ cm$

10. $56\ cm$

11. $52\ cm$

CHAPTER 8: PERIMETER AND AREA SOLUTIONS

8B CIRCUMFERENCE OF A CIRCLE

The perimeter of a circle is called the **circumference**. We can calculate the circumference of a circle by using the formula $C = 2\pi r$, where r is the radius. We can also use $C = \pi d$, where d is the diameter.

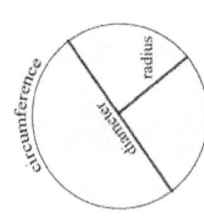

EXERCISE 8B

Find the circumference of each of the following circles.

1. 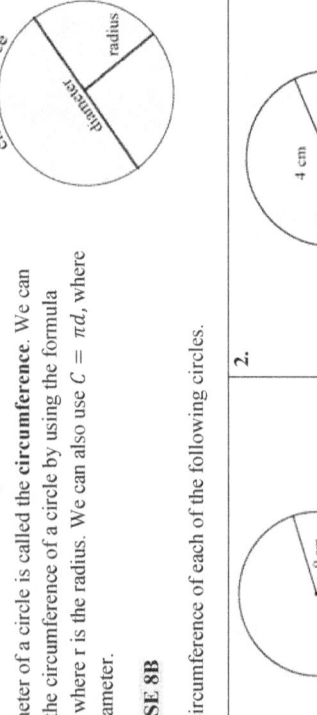 Using $C = 2\pi r$, $C = 2 \times \pi \times 9 = 56.55\ cm$	2. 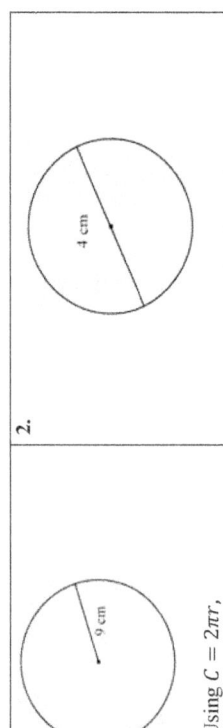 Using $C = \pi d$, $C = \pi \times 4 = 12.57\ cm$
3. 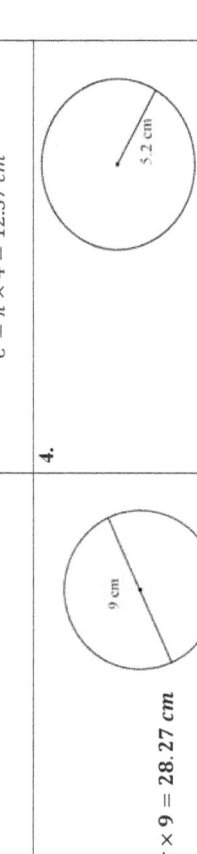 $C = \pi \times 9 = 28.27\ cm$	4. 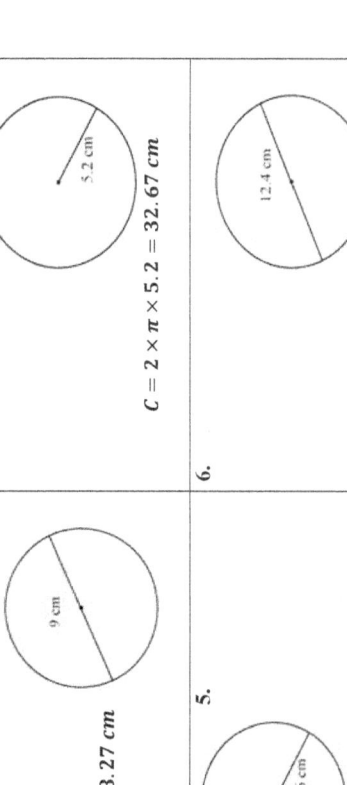 $C = 2 \times \pi \times 5.2 = 32.67\ cm$
5. $C = 2 \times \pi \times 3.6 = 22.62\ cm$	6. $C = \pi \times 12.4 = 38.96\ cm$
7. 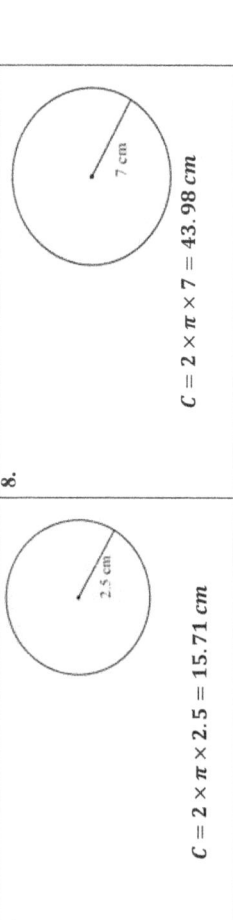 $C = 2 \times \pi \times 2.5 = 15.71\ cm$	8. $C = 2 \times \pi \times 7 = 43.98\ cm$

8C ARC LENGTH AND PERIMETER OF A SECTOR

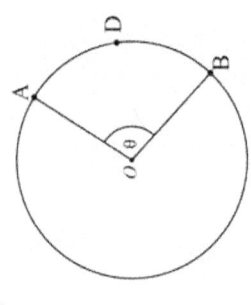

The boundaries of a sector of a circle A are formed by the two radii (OA and OB) and the arc (ADB).
A sector of a circle is just like a slice of pizza.

If we know the length of the radius and the angle subtended at the centre, we can find the length of arc and perimeter of the sector.

The arc length ADB can be found by using the formula

$$l = \frac{\theta}{360} \times 2\pi r,$$

where θ, measured in degrees, is the angle subtended by the arc ADB at the centre of the circle.

To find the perimeter of a sector of a circle, we need to add twice the length of the radius to the length of arc.

Perimeter of a sector is given by $\quad P = \frac{\theta}{360} \times 2\pi r + 2r\ \ or\ P = l + 2r.$

EXAMPLE

For the following sectors, find (i) arc length (ii) the perimeter.

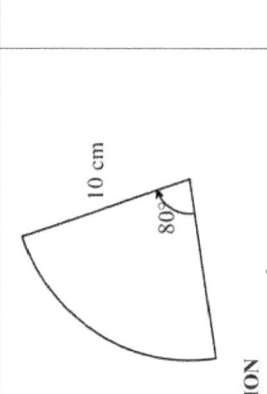

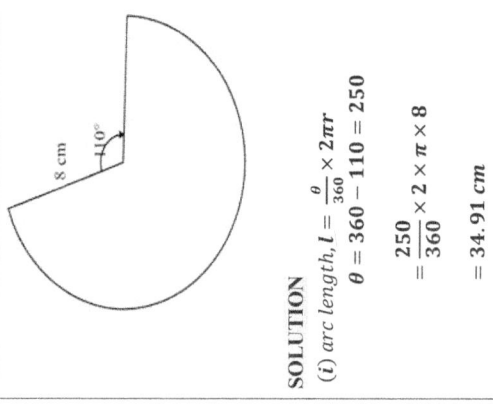

SOLUTION
(i) $arc\ length, l = \frac{\theta}{360} \times 2\pi r$

$= \frac{80}{360} \times 2 \times \pi \times 10$

$= 13.96\ cm$

(ii) Perimeter of minor sector OAB
$= l + 2r$
$= 13.96 + 2 \times 10$
$= 33.96\ cm$

SOLUTION
(i) $arc\ length, l = \frac{\theta}{360} \times 2\pi r$

$\theta = 360 - 110 = 250$

$= \frac{250}{360} \times 2 \times \pi \times 8$

$= 34.91\ cm$

(ii) Perimeter of minor sector OAB
$= l + 2r$
$= 34.91 + 2 \times 8$
$= 50.91\ cm$

CHAPTER 8 : PERIMETER AND AREA SOLUTIONS

EXERCISE 8C

For the following sectors, find (i) arc length (ii) the perimeter.

1.

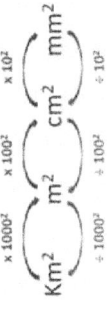

8 cm, 108°

(i) $arc\ length\ l = \frac{108}{360} \times 2 \times \pi \times 8$
$= 15.08\ cm$

(ii) Perimeter of minor sector OAB
$= 15.08 + 2 \times 8$
$= 31.08\ cm$

2.

5.2 cm

(i) $l = \frac{270}{360} \times 2 \times \pi \times 5.2 = 24.50\ cm$

(ii) Perimeter of minor sector OAB
$= 24.50 + 2 \times 5.2$
$= 34.90\ cm$

3.

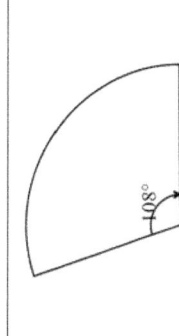

7.2 cm, 81°

(i) $arc\ length\ l = \frac{81}{360} \times 2 \times \pi \times 7.2$
$= 10.18\ cm$

(ii) Perimeter of minor sector OAB
$= 10.18 + 2 \times 7.2$
$= 24.58\ cm$

4.

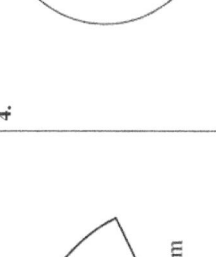

9 cm, 70°

(i) $l = \frac{290}{360} \times 2 \times \pi \times 9 = 45.55\ cm$

(ii) Perimeter of minor sector OAB
$= 45.55 + 2 \times 9$
$= 63.55\ cm$

8D CONVERTING AREA UNITS

Area is simply the amount of surface within a particular boundary. Area is measured in square units such as cm^2, km^2 etc. Because of the complexity of various units used in different situations, we have to be able to convert one unit to the other. The examples below will illustrate the whole process.

As area consists of square units, we have to square all our lengths.

$Km^2 \xrightarrow{\times 1000^2} m^2 \xrightarrow{\times 100^2} cm^2 \xrightarrow{\times 10^2} mm^2$
$\xleftarrow{\div 1000^2} \xleftarrow{\div 100^2} \xleftarrow{\div 10^2}$

CLASS ACTIVITY

Complete the following conversions.

$6\ m^2 \to \ldots\ cm^2$ $6\ m^2 = 6 \times 100^2$ $= 60000\ cm^2$	$1500\ mm^2 \to \ldots\ cm^2$ $1500\ mm^2 = 1500 \div 10^2$ $= 15\ cm^2$
$5\ km^2 \to \ldots\ m^2$ $\mathbf{5 \times 1000^2 = 5\ 000\ 000\ m^2}$	$1200\ cm^2 \to \ldots\ m^2$ $\mathbf{1200 \div 100^2 = 0.12\ m^2}$
$13\ m^2 \to \ldots\ cm^2$ $\mathbf{13 \times 100^2 = 130\ 000\ cm^2}$	$12000\ mm^2 \to \ldots\ cm^2$ $\mathbf{12000 \div 10^2 = 120\ cm^2}$
$500000\ m^2 \to \ldots\ km^2$ $\mathbf{500000 \div 1000^2 = 0.5\ km^2}$	$7\ ha \to \ldots\ m^2$ (1 ha = 10000 m^2) $\mathbf{7 \times 10000 = 70000\ m^2}$
$1400000\ mm^2 \to \ldots\ m^2$ $\mathbf{1400000 \div 10^2 \div 100^2 = 1.4\ m^2}$	$400\ ha \to \ldots\ km^2$ $\mathbf{400ha = 400 \times 10000 = 4000000\ m^2}$ $\mathbf{4\ 000\ 000 \div 1000^2 = 4\ km^2}$

8E AREA FORMULAE

The table below shows different shapes and their respective area formula.

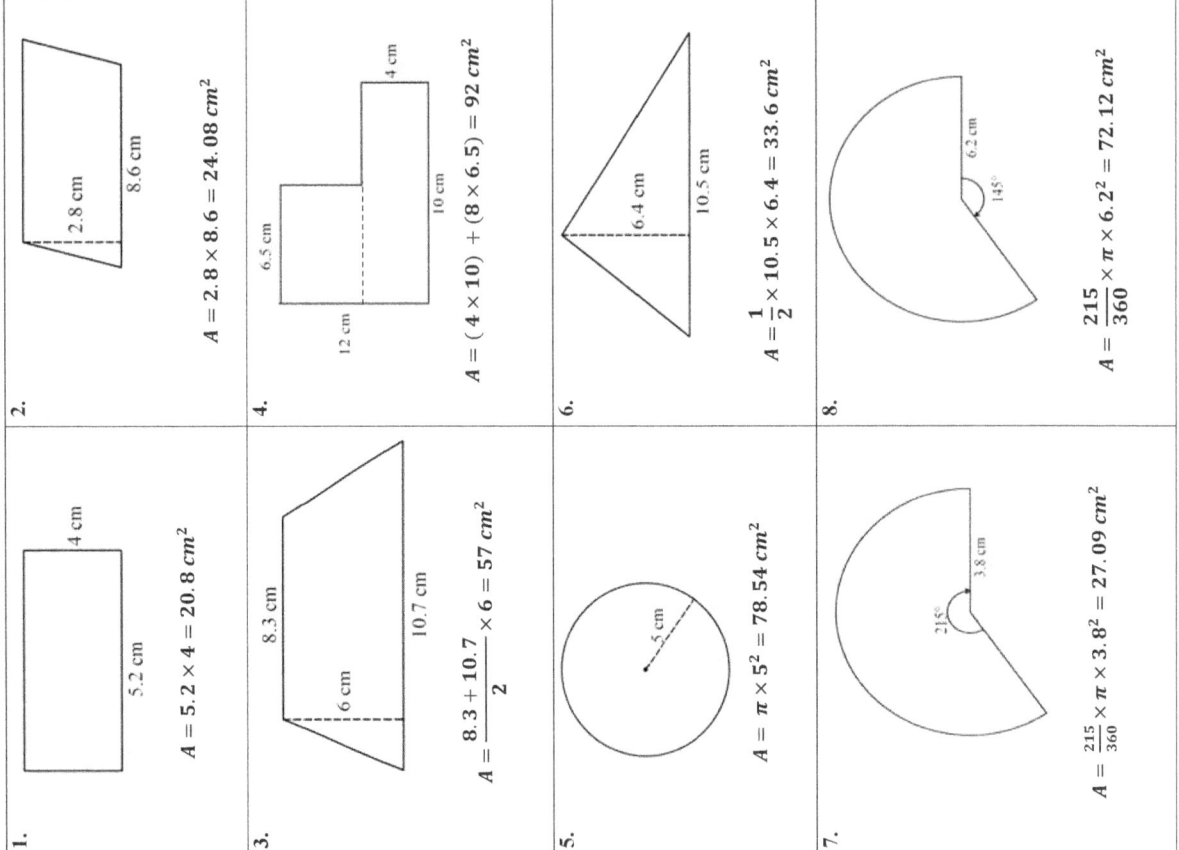

EXERCISE 8E

Find the area of each shape.

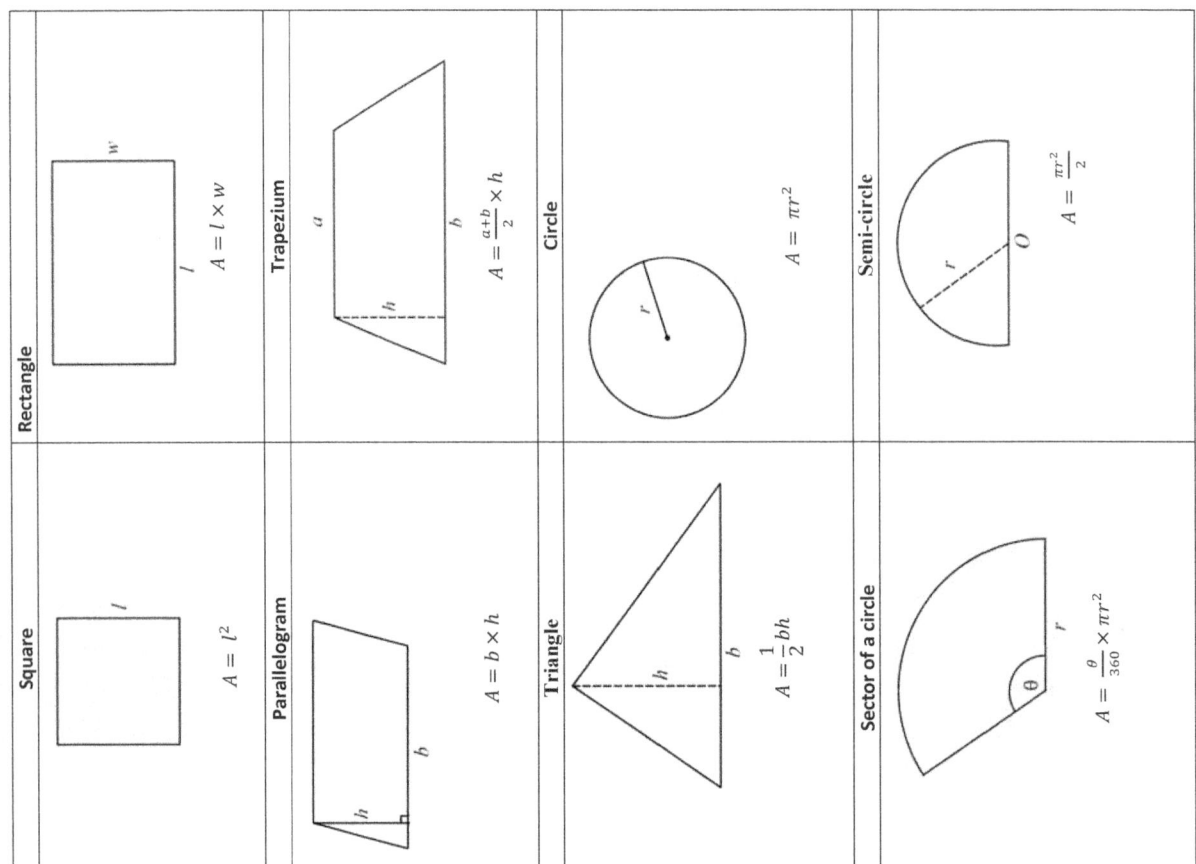

1. $A = 5.2 \times 4 = 20.8\ cm^2$

2. $A = 2.8 \times 8.6 = 24.08\ cm^2$

3. $A = \dfrac{8.3 + 10.7}{2} \times 6 = 57\ cm^2$

4. $A = (4 \times 10) + (8 \times 6.5) = 92\ cm^2$

5. $A = \pi \times 5^2 = 78.54\ cm^2$

6. $A = \dfrac{1}{2} \times 10.5 \times 6.4 = 33.6\ cm^2$

7. $A = \dfrac{215}{360} \times \pi \times 3.8^2 = 27.09\ cm^2$

8. $A = \dfrac{215}{360} \times \pi \times 6.2^2 = 72.12\ cm^2$

8F   FINDING MISSING LENGTH AND RADIUS

In this part of the chapter we are going to find the length of missing sides in different shapes such as square, rectangle, triangle and so on given their respective areas. In addition, another objective would be to find the radius or circumference of a circle given its area.

EXAMPLES

1. A circle has an area of 25 cm^2. Find the radius of the circle.
SOLUTION

$$\pi r^2 = 25$$
$$r^2 = \frac{25}{\pi}$$
$$r = \sqrt{\frac{25}{\pi}} = 2.82 \ cm$$

2. A square has a perimeter of 40 cm. Find the area of the square.
SOLUTION

One side of the square
$= 40 \div 4 = 10 \ cm$
Area of square $= 10 \times 10 = 100 \ cm^2$

3. A circle has a circumference of 24π cm. Find the area of the circle.
SOLUTION

$$2\pi r = 24\pi$$
$$r = 12$$
$$A = \pi(12)^2 = 452.39 \ cm^2$$

4. A triangle has a base length 12 cm and an area of 64.8 cm^2. Find the height of the triangle.
SOLUTION

$$\frac{1}{2}bh = 64.8$$
$$\frac{1}{2} \times 12 \times h = 64.8$$
$$6h = 64.8$$
$$\therefore h = 10.8 \ cm$$

EXERCISE 8F

1. A circle has an area of 36 cm^2. Find the radius of the circle.

$$\pi r^2 = 36$$
$$r^2 = \frac{36}{\pi}$$
$$r = \sqrt{\frac{36}{\pi}} = 3.39 \ cm$$

2. A rectangle has a length of 11 cm. Given that the perimeter of the rectangle is 40 cm, find the width of the rectangle.

$$width = (40 - 2 \times 11) \div 2 = 9 \ cm$$

3. A circle has a circumference of 45.87 cm. Find the area of the circle.

$$2\pi r = 45.87$$
$$r = 7.3 \ cm$$
$$A = \pi(7.3)^2 = 167.42 \ cm^2$$

4. A circle has an area of 36π cm^2. Find the perimeter of the circle.

$$\pi r^2 = 36\pi$$
$$r^2 = 36 \therefore r = 6$$
$$C = 2\pi r = 2 \times \pi \times 6 = 37.70 \ cm$$

Find the shaded area of each of the following shapes. Give your answer to the nearest whole number.

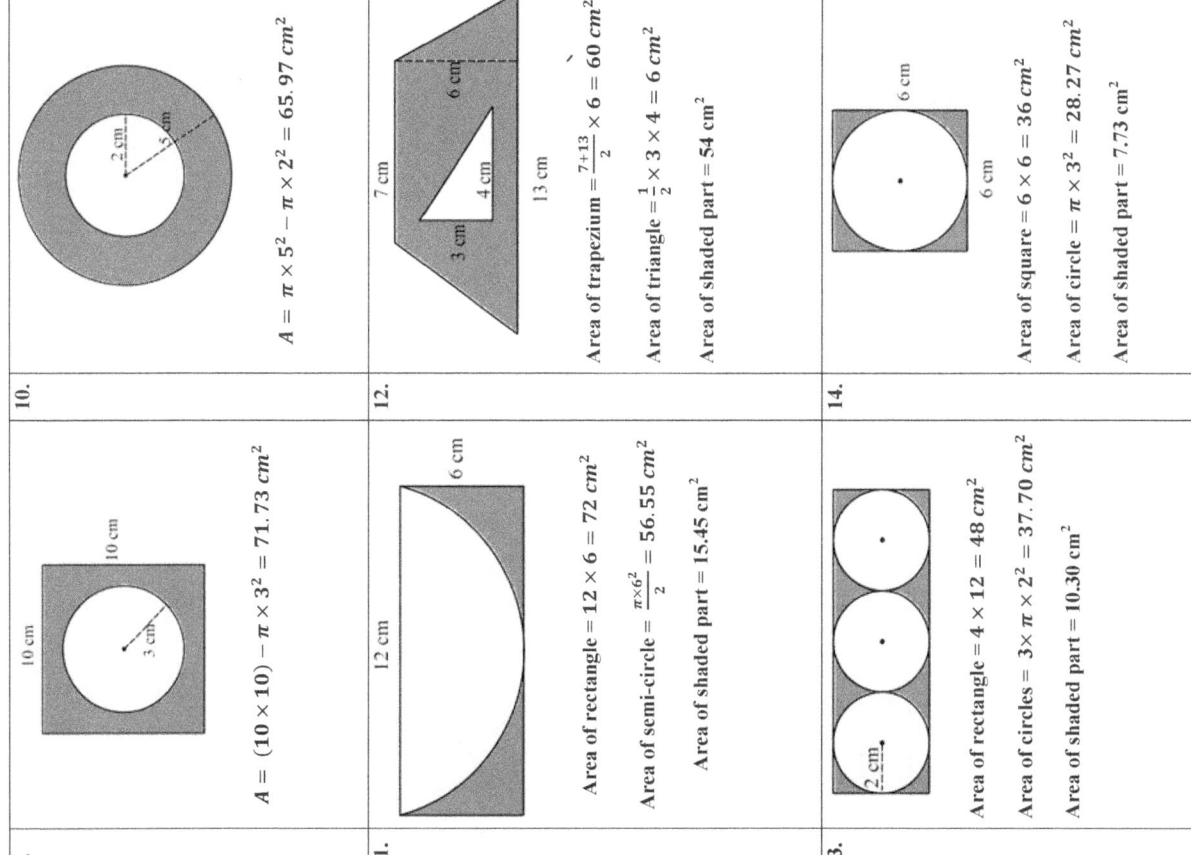

9.
$A = (10 \times 10) - \pi \times 3^2 = 71.73 \ cm^2$

10.
$A = \pi \times 5^2 - \pi \times 2^2 = 65.97 \ cm^2$

11.
Area of rectangle $= 12 \times 6 = 72 \ cm^2$
Area of semi-circle $= \frac{\pi \times 6^2}{2} = 56.55 \ cm^2$
Area of shaded part $= 15.45 \ cm^2$

12.
Area of trapezium $= \frac{7+13}{2} \times 6 = 60 \ cm^2$
Area of triangle $= \frac{1}{2} \times 3 \times 4 = 6 \ cm^2$
Area of shaded part $= 54 \ cm^2$

13.
Area of rectangle $= 4 \times 12 = 48 \ cm^2$
Area of circles $= 3 \times \pi \times 2^2 = 37.70 \ cm^2$
Area of shaded part $= 10.30 \ cm^2$

14.
Area of square $= 6 \times 6 = 36 \ cm^2$
Area of circle $= \pi \times 3^2 = 28.27 \ cm^2$
Area of shaded part $= 7.73 \ cm^2$

CHAPTER 8 : PERIMETER AND AREA SOLUTIONS

5. A square has a perimeter of 28 cm. Find the area of the square.

One side of the square
$= 28 \div 4 = 7\ cm$
Area of square $= 7 \times 7 = 49\ cm^2$

6. A trapezium has its pair of parallel sides as 12 cm and 8 cm respectively. Determine the height of the trapezium if it has an area of 96 cm².

$$\frac{12+8}{2} \times h = 96$$
$$10h = 96$$
$$\therefore h = 9.6\ cm$$

7. A triangle has a base length 12.4 cm and an area of 62 cm². Find the height of the triangle.

$$\frac{12.4 \times h}{2} = 62$$
$$12.4h = 124$$
$$h = 10\ cm$$

8. A circle having diameter 6 cm has the same area as a square of side length x cm. Determine the value of x.

Area of circle $= \pi \times 3^2 = 28.27\ cm^2$
Area of square $= x^2 = 28.27$
$x = 5.3\ cm$

9. The following shapes have the same area. Determine the value of the pronumerals x, y and z.

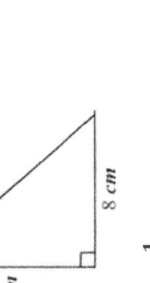

6 cm

6 cm

$A = 6 \times 6 = 36\ cm^2$

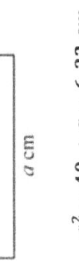

x cm

10 cm

$10x = 36 \therefore x = 3.6\ cm$

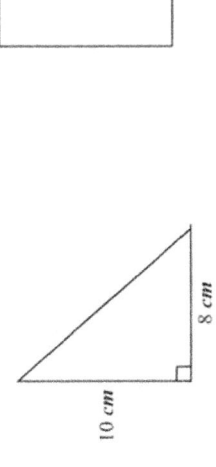

y cm

$\pi \times y^2 = 36 \therefore y = 3.39\ cm$

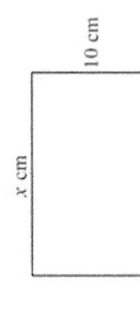

4 cm

6cm

z cm

$\frac{4+6}{2} \times z = 36 \therefore z = 7.2\ cm$

MATHEMATICS APPLICATIONS UNIT 1

10. The following 3 shapes have the same area. Determine the missing sides marked with the letters a and x.

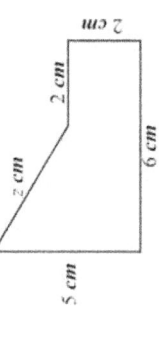

a cm

a cm

$a^2 = 40 \therefore a = 6.32\ cm$

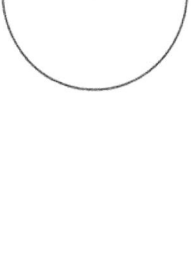

10 cm

8 cm

$A = \frac{1}{2} \times 10 \times 8 = 40\ cm^2$

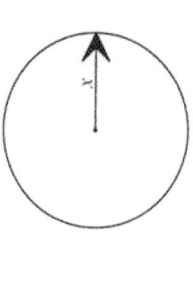

6 cm

x°

$$\frac{x}{360} \times \pi \times 6^2 = 40$$
$$\therefore x = \frac{40 \times 360}{36\pi} = 127.32°$$

11. The following shapes have the same perimeter. Determine the values of x.

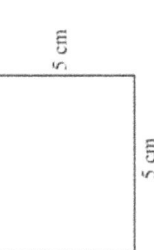

x

$2\pi \times x = 20$
$\therefore x = 3.18\ cm$

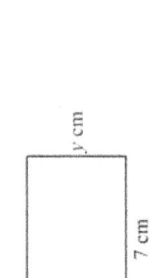

5 cm

5 cm

$P = 5 \times 4 = 20\ cm$

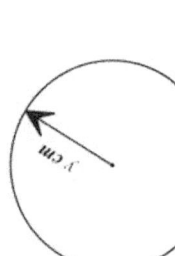

y cm

7 cm

$(7 + y) \times 2 = 20\ cm$
$\therefore y = 3\ cm$

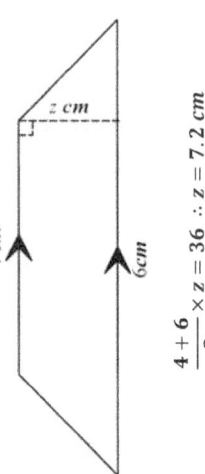

2 cm

z cm

2 cm

6 cm

5 cm

$z = 5\ cm$

CHAPTER 8 : PERIMETER AND AREA SOLUTIONS

8G   APPLICATIONS

1. The diagram shows part of Amy's earring.

It is in the shape of a sector of a circle of radius 6 cm and angle 70°, from which a sector of radius 4 cm and angle 70° has been removed.

Calculate the shaded area.

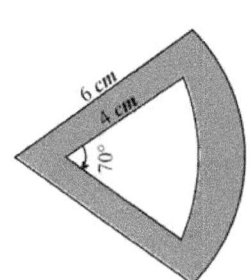

$$A = \left(\frac{70}{360} \times \pi \times 6^2\right) - \left(\frac{70}{360} \times \pi \times 4^2\right) = 12.22 \, cm^2$$

2. ABCD is a level field.
F and E are points on AD such that BF and CE are perpendicular to AD.

BF = 40 m and CE = 52 m.

AF = 16 m, FE = 32 m and ED = 40 m.

(a) Calculate the area of the field.

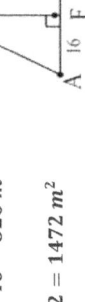

$Area\ of\ \triangle ABF = = \frac{1}{2} \times 16 \times 40 = 320 \, m^2$

$Area\ of\ BCEF = = \frac{40+52}{2} \times 32 = 1472 \, m^2$

$Area\ of\ \triangle CED = = \frac{1}{2} \times 52 \times 40 = 1040 \, m^2$

$Total\ area = 320 + 1472 + 1040 = 2832 \, m^2$

(b) Calculate the length of BC.

$BC = \sqrt{32^2 + 12^2} = 34.18 \, m$

3. AD and BC are arcs of circles with centre O.

A is a point on OB, and D is a point on OC.

OA = 10 cm and AB = 15 cm.

∠AOD = 130°.

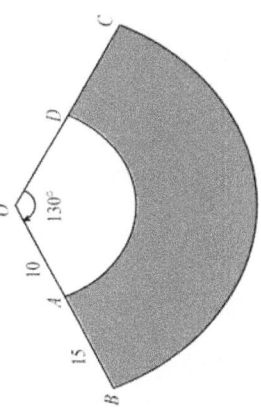

(a) Calculate the perimeter of the shaded shape ABCD.

$Arc\ AD = \frac{130}{360} \times 2 \times \pi \times 10 = 22.69$

$Arc\ BC = \frac{130}{360} \times 2 \times \pi \times 25 = 56.72$

$Perimeter\ of\ ABCD = 22.69 + 56.72 + 15 + 15 = 109.41 \, cm$

(b) Calculate the area of the shaded shape ABCD.

$A = \left(\frac{130}{360} \times \pi \times 25^2\right) - \left(\frac{130}{360} \times \pi \times 10^2\right) = 595.59 \, cm^2$

(c) The shape ABCD is used to make a lampshade by joining AB and DC.

Calculate the radius, r cm, of the circular top of the lampshade.

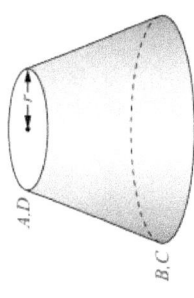

$circumference\ of\ top\ of\ the\ lampshade = 22.69$

$2\pi \times r = 22.69$

$r = \frac{22.69}{2\pi} = 3.61 \, cm$

CHAPTER 8 : PERIMETER AND AREA SOLUTIONS

4. The Ryan's family recently bought a beach house in Rockingham. The house was built in the 1990's and as a result needs some renovation. As a priority, the family decided to tile all the four bedrooms as the carpet was old and their younger son being asthmatic was allergic to carpet. They had a few quotes and their best rate was $21.70 m^2 from ABC tilers.

ABC TILERS CO.
$21.70 m^2

(a) Calculate how much it would cost them in total to tile all the four bedrooms.

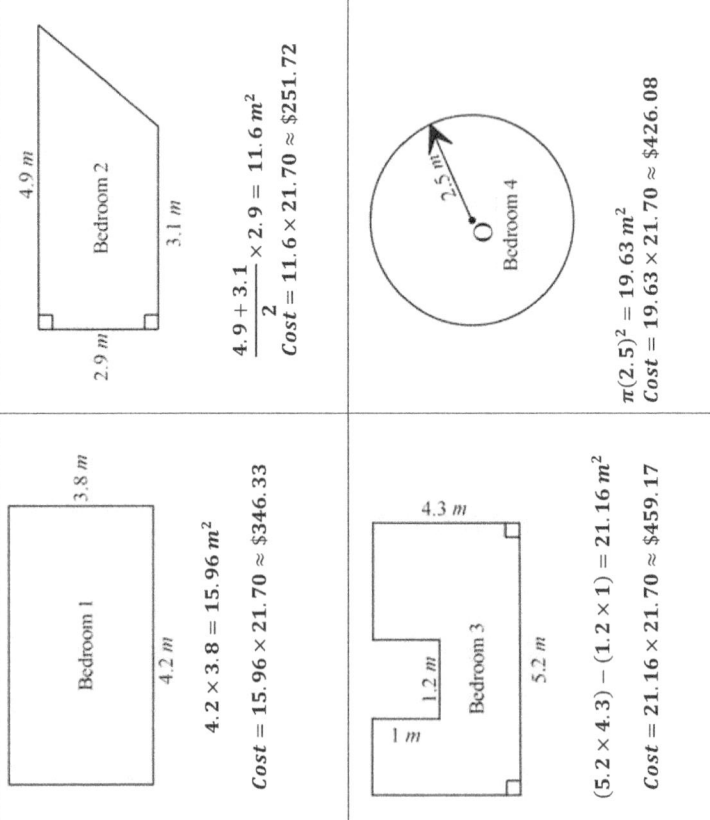

Bedroom 1: 4.2 m × 3.8 m

$4.2 \times 3.8 = 15.96\ m^2$

$Cost = 15.96 \times 21.70 \approx \$346.33$

Bedroom 2: 4.9 m, 2.9 m, 3.1 m

$\dfrac{4.9 + 3.1}{2} \times 2.9 = 11.6\ m^2$

$Cost = 11.6 \times 21.70 \approx \$251.72$

Bedroom 3: 4.3 m, 5.2 m, 1.2 m, 1 m

$(5.2 \times 4.3) - (1.2 \times 1) = 21.16\ m^2$

$Cost = 21.16 \times 21.70 \approx \$459.17$

Bedroom 4: radius 2.5 m

$\pi(2.5)^2 = 19.63\ m^2$

$Cost = 19.63 \times 21.70 \approx \$426.08$

$Total\ Cost = 346.33 + 251.72 + 459.17 + 426.08 = \$1483.30$

(b) Bedroom 4 being harder to tile, the Ryan's family was charged 5% extra for tiling that room. Determine the increase in cost.

$Increase\ in\ Cost = 426.08 \times 0.05 \approx \$21.30$

MATHEMATICS APPLICATIONS UNIT 1

5. Delta Pools are currently advertising 4 of its latest pools. The company is offering the Value collection pavers which are ideal for first home buyers or people looking to add a little beauty to their investment property.
Determine the cost to the nearest dollar of paving around each of the following pool areas.

DELTA POOLS
Huge Discount
$24.80 m^2

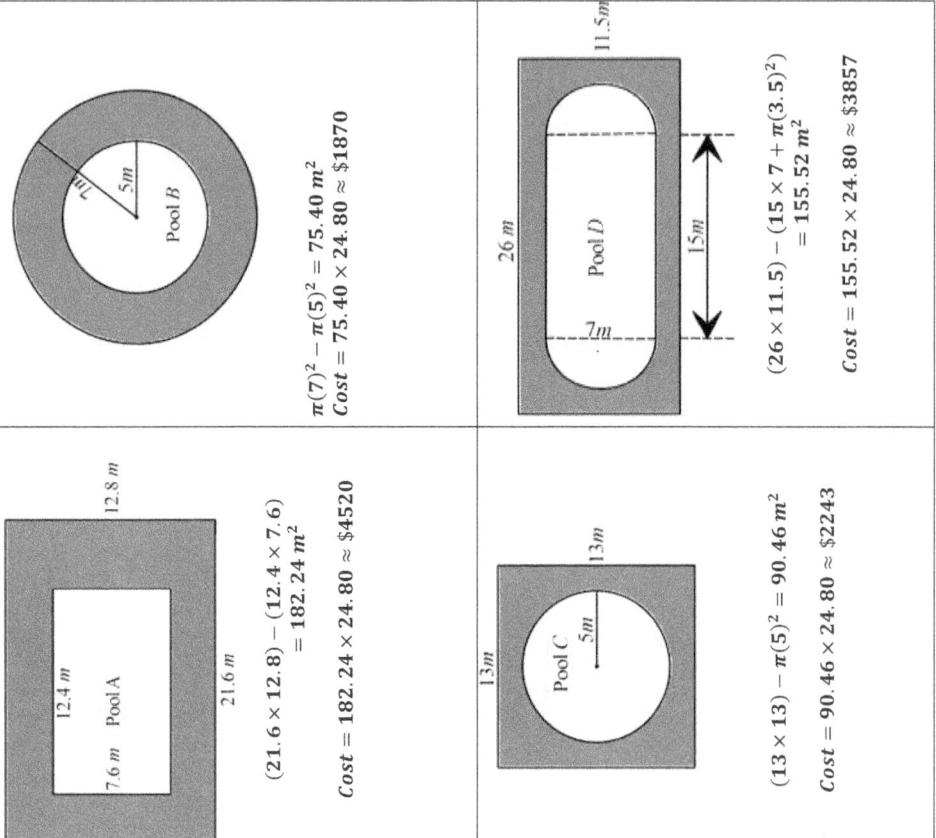

Pool A: 12.4 m × 7.6 m inner, 21.6 m × 12.8 m outer

$(21.6 \times 12.8) - (12.4 \times 7.6) = 182.24\ m^2$

$Cost = 182.24 \times 24.80 \approx \$4520$

Pool B: inner radius 5 m, outer radius 7 m

$\pi(7)^2 - \pi(5)^2 = 75.40\ m^2$

$Cost = 75.40 \times 24.80 \approx \$1870$

Pool C: 13 m × 13 m square, circle radius 5 m

$(13 \times 13) - \pi(5)^2 = 90.46\ m^2$

$Cost = 90.46 \times 24.80 \approx \$2243$

Pool D: 26 m × 11.5 m outer, 7 m width, 15 m length inner

$(26 \times 11.5) - (15 \times 7 + \pi(3.5)^2) = 155.52\ m^2$

$Cost = 155.52 \times 24.80 \approx \$3857$

CHAPTER 9

SURFACE AREA AND VOLUME

9A SURFACE AREA AND VOLUME FORMULAE

Name of solid	Figure	Total surface Area	Volume
Cube		$A = 6x^2$	$V = x^3$
Rectangular prism		$A = 2(lw + wh + hl)$	$V = l \times w \times h$
Triangular Prism		$2 \times$ area of Δ + area of 3 rectangles	$V = \frac{1}{2}bh \times l$ or $V = $ Area of $\Delta \times l$
Cylinder		$A = 2\pi r(r + h)$ or $A = 2\pi r^2 + 2\pi rh$	$V = \pi r^2 h$
Sphere		$A = 4\pi r^2$	$V = \frac{4}{3}\pi r^3$
Cone		$A = \pi r(r + l)$ or $A = \pi r^2 + \pi r l$	$V = \frac{1}{3}\pi r^2 h$
Pyramid		Add base area + area of all Δs	$V = \frac{1}{3}($Area of base$) \times h$

EXERCISE 9A

Find the total surface area and volume of each of the following solids. All units are in cm.

	Solids	Total surface area	Volume
1.	(cylinder, r=3, h=10)	$A = 2\pi r^2 + 2\pi rh$ $2 \times \pi \times 3^2 + 2 \times \pi \times 3 \times 10$ $= 245.04\ cm^2$	$V = \pi r^2 h$ $V = \pi \times 3^2 \times 10$ $= 282.74\ cm^3$
2.	(cylinder, d=21, h=8)	$A = 2\pi r^2 + 2\pi rh$ $2 \times \pi \times 10.5^2 + 2 \times \pi \times 10.5 \times 8$ $= 1220.51\ cm^2$	$V = \pi r^2 h$ $V = \pi \times 10.5^2 \times 8$ $= 2770.88\ cm^3$
3.	(cylinder, r=3.5, h=15)	$2 \times \pi \times 3.5^2 + 2 \times \pi \times 3.5 \times 15$ $= 406.84\ cm^2$	$V = \pi \times 3.5^2 \times 15$ $= 577.27\ cm^3$
4.	(cylinder, r=2.5, h=8)	$2 \times \pi \times 2.5^2 + 2 \times \pi \times 2.5 \times 8$ $= 164.93\ cm^2$	$V = \pi \times 2.5^2 \times 8$ $= 157.08\ cm^3$

CHAPTER 9 : SURFACE AREA AND VOLUME SOLUTIONS

	Solids	Total surface area	Volume
5.	(rectangular prism 7.5 × 3 × 4)	$A = 2(lw + wh + hl)$ $2(7.5 \times 3 + 3 \times 4 + 4 \times 7.5)$ $= 129\ cm^2$	$V = l \times w \times h$ $V = 7.5 \times 3 \times 4 = 90\ cm^3$
6.	(cube 8)	$A = 6x^2$ $A = 6 \times 8^2 = 384\ cm^2$	$V = x^3$ $V = 8^3 = 512\ cm^3$
7.	(rectangular prism 12 × 5 × 6)	$A = 2(lw + wh + hl)$ $2(12 \times 5 + 5 \times 6 + 6 \times 12)$ $= 324\ cm^2$	$V = l \times w \times h$ $V = 12 \times 5 \times 6$ $= 360\ cm^3$
8.	(cube 9)	$A = 6x^2$ $A = 6 \times 9^2 = 486\ cm^2$	$V = x^3$ $V = 9^3 = 729\ cm^3$
9.	(rectangular prism 10 × 4 × 7)	$A = 2(lw + wh + hl)$ $2(10 \times 4 + 4 \times 7 + 7 \times 10)$ $= 276\ cm^2$	$V = l \times w \times h$ $V = 10 \times 4 \times 7$ $= 280\ cm^3$
10.	(cube 6.1)	$A = 6x^2$ $A = 6 \times 6.1^2$ $= 223.26\ cm^2$	$V = x^3$ $V = 6.1^3 = 226.981\ cm^3$

MATHEMATICS APPLICATIONS UNIT 1

	Solids	Total surface area	Volume
11.	(cone r=6, h=8)	Use Pythagoras to find the slant height. $l^2 = 6^2 + 8^2$ $\therefore l = 10$ $A = \pi r^2 + \pi r l$ $A = \pi \times 6^2 + \pi \times 6 \times 10$ $= 301.59\ cm^2$	$V = \frac{1}{3}\pi r^2 h$ $V = \frac{1}{3} \times \pi \times 6^2 \times 8$ $= 301.59\ cm^3$
12.	(cone l=13, h=12)	Using Pythagoras to find the radius $r = \sqrt{13^2 - 12^2} = 5$ $A = \pi r^2 + \pi r l$ $A = \pi \times 5^2 + \pi \times 5 \times 13$ $= 282.74\ cm^2$	$V = \frac{1}{3}\pi r^2 h$ $V = \frac{1}{3} \times \pi \times 5^2 \times 12$ $= 314.16\ cm^3$
13.	(cone r=5, h=24)	Using Pythagoras to find the slant height $l = \sqrt{24^2 + 12^2} = 24.5$ $A = \pi r^2 + \pi r l$ $A = \pi \times 5^2 + \pi \times 5 \times 24.5$ $= 463.38\ cm^2$	$V = \frac{1}{3}\pi r^2 h$ $V = \frac{1}{3} \times \pi \times 5^2 \times 24$ $= 628.32\ cm^3$
14.	(cone l=15, h=12)	Using Pythagoras to find the radius $r = \sqrt{15^2 - 12^2} = 9$ $A = \pi r^2 + \pi r l$ $A = \pi \times 9^2 + \pi \times 9 \times 15$ $= 678.58\ cm^2$	$V = \frac{1}{3}\pi r^2 h$ $V = \frac{1}{3} \times \pi \times 9^2 \times 12$ $= 1017.88\ cm^3$

CHAPTER 9 : SURFACE AREA AND VOLUME SOLUTIONS

	Solids	Total surface area	Volume
15.	(sphere, r=4)	$A = 4\pi r^2$ $A = 4 \times \pi \times 4^2$ $= 201.06\ cm^2$	$V = \dfrac{4}{3}\pi r^3$ $V = \dfrac{4}{3} \times \pi \times 4^3$ $= 268.08\ cm^3$
16.	(hemisphere, r=3.6)	$A = 3\pi r^2$ $A = 3 \times \pi \times 3.6^2$ $= 122.15\ cm^2$	$V = \dfrac{2}{3}\pi r^3$ $V = \dfrac{2}{3} \times \pi \times 3.6^3$ $= 97.72\ cm^3$
17.	(sphere, r=4.5)	$A = 4\pi r^2$ $A = 4 \times \pi \times 4.5^2$ $= 254.47\ cm^2$	$V = \dfrac{4}{3}\pi r^3$ $V = \dfrac{4}{3} \times \pi \times 4.5^3$ $= 381.70\ cm^3$
18.	(hemisphere, r=5)	$A = 3\pi r^2$ $A = 3 \times \pi \times 5^2$ $= 235.62\ cm^2$	$V = \dfrac{2}{3}\pi r^3$ $V = \dfrac{2}{3} \times \pi \times 5^3$ $= 261.80\ cm^3$
19.	(hemisphere, r=3)	$A = 3\pi r^2$ $A = 3 \times \pi \times 3^2$ $= 84.82\ cm^2$	$V = \dfrac{2}{3}\pi r^3$ $V = \dfrac{2}{3} \times \pi \times 3^3$ $= 56.55\ cm^3$
20.	(sphere, r=6.2)	$A = 4\pi r^2$ $A = 4 \times \pi \times 6.2^2$ $= 483.05\ cm^2$	$V = \dfrac{4}{3}\pi r^3$ $V = \dfrac{4}{3} \times \pi \times 6.2^3$ $= 998.31\ cm^3$

	Solids	Total surface area	Volume
21.	(triangular prism 5,4,6,18)	$2 \times \text{area of } \Delta$ $= 2 \times \dfrac{1}{2} \times 6 \times 4 = 24$ Area of 3 rectangles $= 6 \times 18 + 18 \times 5 + 18 \times 5$ $= 288\ cm^2$ Total surface area $= 312\ cm^2$	$V = \text{Area of } \Delta \times l$ $V = \dfrac{1}{2} \times 6 \times 4 \times 18$ $= 216\ cm^3$
22.	(triangular prism 10,6,16,30)	$TSA = \left(2 \times \dfrac{1}{2} \times 16 \times 6\right) +$ $(16 \times 30 + 30 \times 10 + 30 \times 10)$ $= 1176\ cm^2$	$V = \dfrac{1}{2} \times 16 \times 6 \times 30$ $= 1440\ cm^3$
23.	(triangular prism 3,10,4,x)	$x = \sqrt{3^2 + 4^2} = 5$ $TSA = \left(2 \times \dfrac{1}{2} \times 3 \times 4\right) +$ $(3 \times 10 + 4 \times 10 + 5 \times 10)$ $= 132\ cm^2$	$V = \dfrac{1}{2} \times 3 \times 4 \times 10$ $= 60\ cm^3$
24.	(triangular prism 25,24,14,50)	$TSA = \left(2 \times \dfrac{1}{2} \times 14 \times 24\right) +$ $(14 \times 50 + 50 \times 25 + 50 \times 25)$ $= 3536\ cm^2$	$V = \dfrac{1}{2} \times 14 \times 24 \times 50$ $= 8400\ cm^3$
25.	(triangular prism 9,15,20)	$x = \sqrt{15^2 - 9^2} = 12$ $TSA = \left(2 \times \dfrac{1}{2} \times 9 \times 12\right) +$ $(12 \times 20 + 9 \times 20 + 15 \times 20)$ $= 828\ cm^2$	$V = \dfrac{1}{2} \times 9 \times 12 \times 20$ $= 1080\ cm^3$

CHAPTER 9: SURFACE AREA AND VOLUME SOLUTIONS

	Solids	Total surface area	Volume
26.	(pyramid with base 10×10, slant 13, height 12)	Add base area + area of all Δs Base area $= 10 \times 10 = 100$ Area of 4 triangles $= 4 \times \frac{1}{2} \times 10 \times 13 = 260$ Total surface area $= 360\ cm^2$	$V = \frac{1}{3}(Area\ of\ base) \times h$ $V = \frac{1}{3} \times 10 \times 10 \times 12$ $= 400\ cm^3$
27.	(pyramid base 12×12, slant 10, height 8)	Base area $= 12 \times 12 = 144$ Area of 4 triangles $= 4 \times \frac{1}{2} \times 12 \times 10 = 240$ $TSA = 144 + 240 = 384\ cm^2$	$V = \frac{1}{3}(Area\ of\ base) \times h$ $V = \frac{1}{3} \times 12 \times 12 \times 8$ $= 384\ cm^3$
28.	(pyramid base 16×16, slant 17, height 15)	Base area $= 16 \times 16 = 256$ Area of 4 triangles $= 4 \times \frac{1}{2} \times 16 \times 17 = 544$ $TSA = 256 + 544 = 800\ cm^2$	$V = \frac{1}{3}(Area\ of\ base) \times h$ $V = \frac{1}{3} \times 16 \times 16 \times 15$ $= 1280\ cm^3$
29.	(pyramid base 18×18, slant 15, height 12)	$x = \sqrt{15^2 - 12^2} = 9$ Base area $= 18 \times 18 = 324$ Area of 4 triangles $= 4 \times \frac{1}{2} \times 18 \times 15 = 540$ $TSA = 324 + 540 = 864\ cm^2$	$V = \frac{1}{3}(Area\ of\ base) \times h$ $V = \frac{1}{3} \times 18 \times 18 \times 12$ $= 1296\ cm^3$
30.	(pyramid base 6×6, 4 cm, 6 cm)	$x = \sqrt{3^2 + 4^2} = 5$ Base area $= 6 \times 6 = 36$ Area of 4 triangles $= 4 \times \frac{1}{2} \times 6 \times 5 = 60$ $TSA = 60 + 36 = 96\ cm^2$	$V = \frac{1}{3}(Area\ of\ base) \times h$ $V = \frac{1}{3} \times 6 \times 6 \times 4$ $= 48\ cm^3$

9B FINDING RADIUS OR SIDE

In this section of the chapter, the volume of different shapes would be given and our aim would be to find the radius or missing sides from cylinders, cones, spheres, cube and so on.

EXAMPLES

In the examples below, we can alternatively make use of the solve facility of the CAS calculator to find the missing values.

1. A sphere has a volume of $400\ cm^3$. Find the radius of the sphere correct to one decimal place.

   SOLUTION
   $$\frac{4}{3}\pi r^3 = 400$$
   $$r^3 = \frac{400 \times 3}{4\pi}$$
   $$r = \sqrt[3]{\frac{1200}{4\pi}} = 4.6\ cm$$

2. A cube has surface area of $150\ cm^2$. Find the volume of the cube.

   SOLUTION
   $$6x^2 = 150$$
   $$x^2 = 25$$
   $$\therefore x = 5$$
   Volume of cube $= 5^3 = 125\ cm^3$

3. A cylinder has a radius of 6 cm and a volume of $1244.07\ cm^3$. Determine the height of the cylinder to the nearest cm.

   SOLUTION
   $\pi r^2 h = 1244.07$
   $\pi \times 6^2 \times h = 1244.07$
   $h = \frac{1244.07}{36\pi} = 11\ cm$

4. A right circular cone has a volume of $442.44\ cm^3$. Given that its perpendicular height is 10 cm, determine the radius of the cone.

   SOLUTION
   $\frac{1}{3}\pi r^2 h = 442.44$
   $\frac{1}{3}\pi \times r^2 \times 10 = 442.44$
   $r^2 = \frac{442.44 \times 3}{10\pi}$
   $r = \sqrt{\frac{442.44 \times 3}{10\pi}} = 6.5\ cm$

5. A cube of side 5 cm and a sphere of radius 6 cm are both melted down to form a cylinder of height 10 cm. Determine the radius of the cylinder.

   SOLUTION
   Volume of cube $= 5^3 = 125\ cm^3$
   Volume of sphere $= \frac{4}{3} \times \pi \times 6^3 = 904.78\ cm^3$
   Total volume = 1029.78
   CAS ($Solve\ (\pi \times r^2 \times 10 = 1029.78, r)$
   $r = 5.73\ cm$

EXERCISE 9B

1. A sphere has a volume of 360 cm³. Find the radius of the sphere correct to one decimal place.

$$\frac{4}{3}\pi r^3 = 360$$
$$r^3 = \frac{360 \times 3}{4\pi}$$
$$r = \sqrt[3]{\frac{1080}{4\pi}} = 4.41\ cm$$

2. A cube has a volume of 216 cm³. Find the total surface area of the cube.

$$x^3 = 216$$
$$x = 6$$
$$TSA = 6(6^2) = 216\ cm^2$$

3. A cylindrical tank of height 40 cm and radius r cm has a capacity of 18 litres. Find the radius to the nearest centimetre. (1 litre = 1000 cm³)

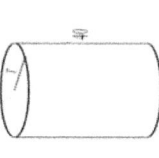

$$\pi \times r^2 \times 40 = 18000$$
$$r^2 = \frac{18000}{40\pi}$$
$$\therefore r = 11.97 \approx 12\ cm$$

4. The cylinder on the right has a radius of 4 cm and height h cm. The volume of the cylinder is 1200 cm³. Find h.

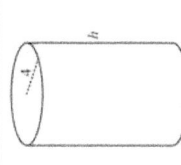

$$\pi \times 4^2 \times h = 1200$$
$$h = \frac{1200}{16\pi}$$
$$\therefore h = 23.87\ cm$$

5. A right circular cone has a volume of 603.19 cm³. Given that its radius is 8 cm, determine the perpendicular height of the cone.

$$\frac{1}{3}\pi r^2 h = 603.19$$
$$\frac{1}{3}\pi \times 8^2 \times h = 603.19$$
$$h = \frac{603.19 \times 3}{64\pi}$$
$$h = 9\ cm$$

6. A cube of side 10 cm is melted down to form a cylinder of height 8 cm. Determine the radius of the cylinder.

Volume of cube $= 10^3 = 1000\ cm^3$

$$\pi \times r^2 \times 8 = 1000$$
$$r = \sqrt{\frac{1000}{8\pi}}$$
$$r = 6.3\ cm$$

7. The following six shapes have the same volume. Determine each of the pronumerals marked x, y, z, w and m.

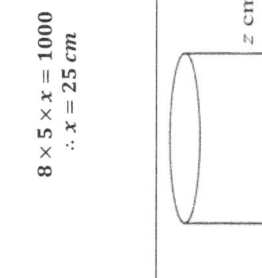

$8 \times 5 \times x = 1000$
$\therefore x = 25\ cm$

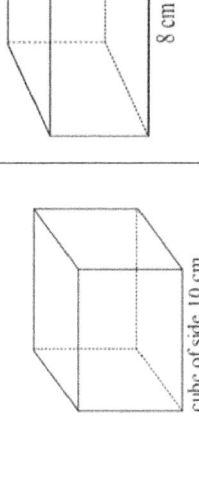

cube of side 10 cm
$Volume = 10^3 = 1000\ cm^3$

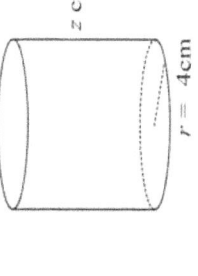

$\frac{1}{3}\pi \times y^2 \times 8 = 1000$
$y = \sqrt{\frac{3000}{8\pi}} = 10.93\ cm$

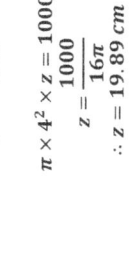

$\pi \times 4^2 \times z = 1000$
$z = \frac{1000}{16\pi}$
$\therefore z = 19.89\ cm$

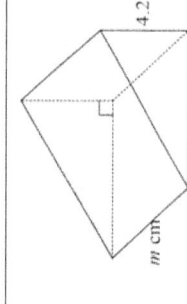

$\frac{4}{3} \times \pi \times w^3 = 1000$
$r^3 = \frac{1000 \times 3}{4\pi}$
$r = \sqrt[3]{\frac{3000}{4\pi}} = 6.20\ cm$

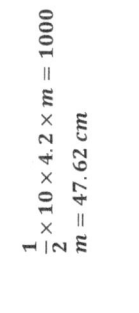

$\frac{1}{2} \times 10 \times 4.2 \times m = 1000$
$m = 47.62\ cm$

CHAPTER 9 : SURFACE AREA AND VOLUME SOLUTIONS

8. The following four shapes have the same total surface area. Determine each of the pronumerals marked a, h and x.

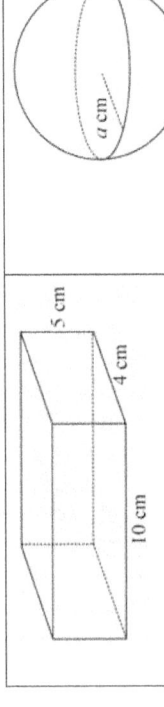

$TSA = 2(10 \times 4 + 4 \times 5 + 5 \times 10)$
$= 220 \, cm^2$

$4 \times \pi \times a^2 = 220$
$y = \sqrt{\dfrac{220}{4\pi}} = 4.18 \, cm$

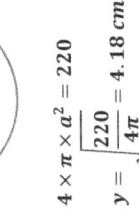

$2 \times \pi \times 3^2 + 2 \times \pi \times 3 \times h = 220$
$6\pi h = 220 - 18\pi$
$h = 8.67 \, cm$

cube of side x cm

$6x^2 = 220 \therefore x = \sqrt{\dfrac{220}{6}} = 6.06 \, cm$

9. The solid below has a total surface area of 162 cm². Determine the value of x.

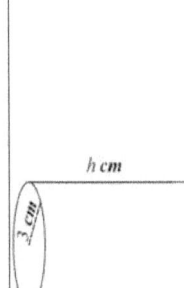

$3x + 6x + 5x + 4x + 2 \times \left(\dfrac{3+6}{2} \times 4\right) = 162$
$18x + 36 = 162$
$18x = 126$
$\therefore x = 7$

9C CAPACITY

Capacity is the volume of liquid that can fit inside a container. For example, the capacity of a carton of milk is the amount of milk that can fit inside the carton.
The standard units for capacity are the litre (L) and millilitre (mL). Note the following:

$1 \, mL = 1 \, cm^3$
$1 \, litre = 1000 \, mL = 1000 cm^3$
$1 \, kL = 1000 L$
$1 ML = 1 000 000 \, L$

CLASS ACTIVITY

Complete the following table

1. 50 cm^3 = **50** mL	2. 4500 mL = **4.5** L
3. 4 L = **4000** mL	4. 3.4 L = **3400** cm^3
5. 7000 cm^3 = **7** L	6. 3.65 kL = **3650** L
7. 33.7 L = **33700** cm^3	8. 3850 L = **3.85** kL
9. 900 mL = **900** cm^3	10. 2.54 L = **2540** cm^3

11. A fish tank is in the shape of a rectangular prism. It is 65cm long, 40 cm wide and 50cm high. Determine the capacity of the tank.

$Volume = 65 \times 40 \times 50 = 130\,000 \, cm^3 = 130 \, L$

12. Find the capacity of the container on the right.

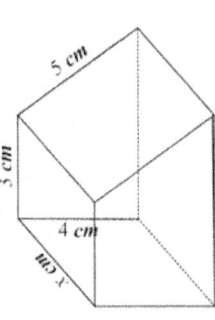

$Volume = \dfrac{2}{3} \times \pi \times 4.2^3 + \pi \times 4.2^2 \times 13$
$= 875.60 \, cm^3$
$= 0.876 \, L$

CHAPTER 9 : SURFACE AREA AND VOLUME SOLUTIONS

13. The following three cylindrical containers are filled with a certain liquid. Rank them according to their capacity, giving your answer in descending order.

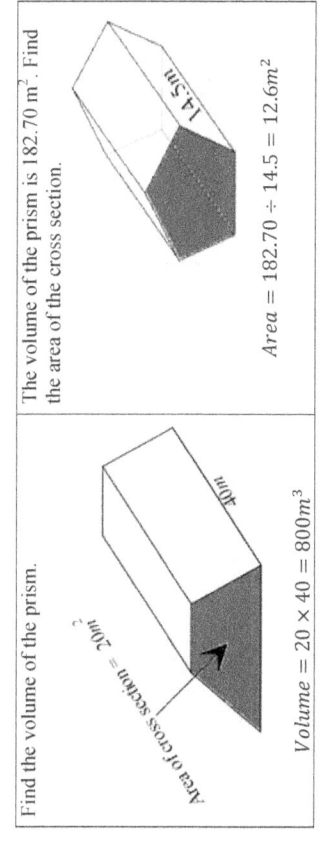

$Volume\ of\ A = \pi \times 4.5^2 \times 12 = 763.41\ cm^3 = 763.41\ mL\ or\ 0.763\ L$
$Volume\ of\ B = \pi \times 3.9^2 \times 14 = 668.97\ cm^3 = 668.97\ mL\ or\ 0.669\ L$
$Volume\ of\ C = \pi \times 4.1^2 \times 13 = 686.53\ cm^3 = 686.53\ mL\ or\ 0.687\ L$
$RANK: A, C, then\ B$

14. Three cartons P, Q or R are filled with milk. Which of them has the greatest capacity? Show your workings.

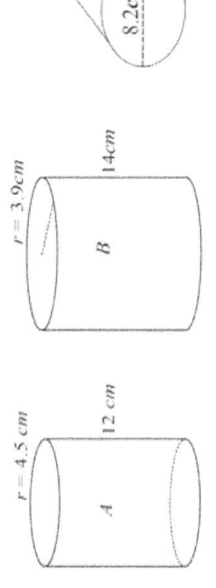

$Volume\ of\ P = 20 \times 15 \times 16 = 4800\ cm^3 = 4.8\ L$
$Volume\ of\ Q = 24 \times 13 \times 14 = 4368\ cm^3 = 4.368\ L$
$Volume\ of\ R = 10.8 \times 12.8 \times 18.6 = 2571.26\ cm^3 = 2.57\ L$
$RANK: P, Q\ then\ R$

9D   VOLUME OF OTHER PRISMS

The volume of a prism is given by the formula:

$$Volume\ of\ prism = Area\ of\ cross\ section \times length$$

EXAMPLES

Find the volume of the prism.

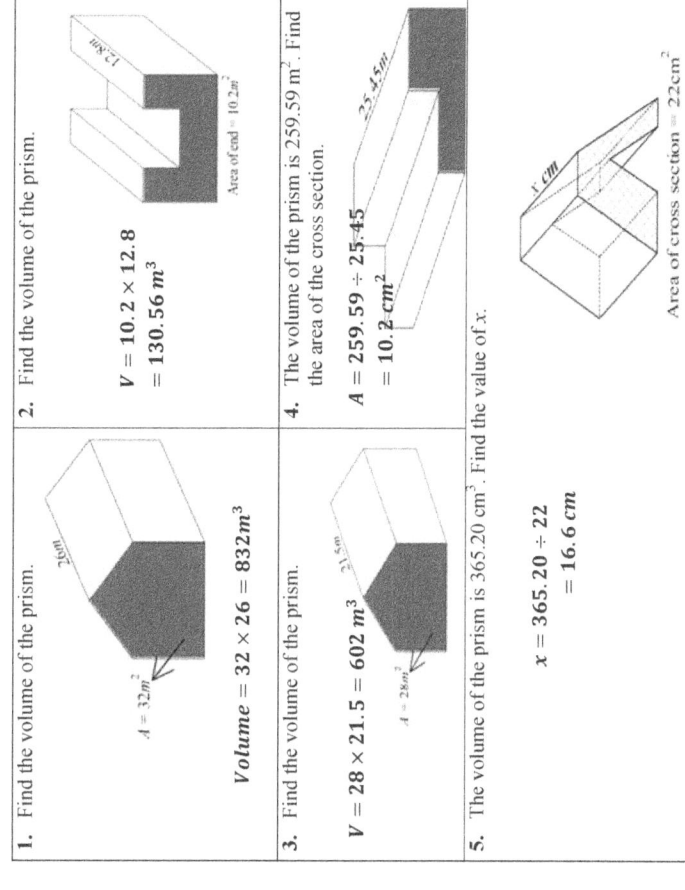

The volume of the prism is 182.70 m². Find the area of the cross section.

$Volume = 20 \times 40 = 800\ m^3$

$Area = 182.70 \div 14.5 = 12.6\ m^2$

EXERCISE 9D

1. Find the volume of the prism.

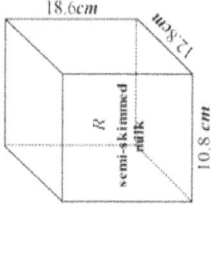

2. Find the volume of the prism.

$V = 10.2 \times 12.8$
$= 130.56\ m^3$

3. Find the volume of the prism.

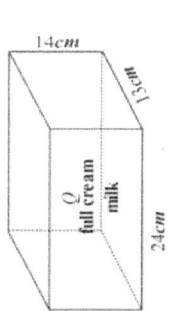

$V = 28 \times 21.5 = 602\ m^3$

4. The volume of the prism is 259.59 m². Find the area of the cross section.

$A = 259.59 \div 25.45$
$= 10.2\ cm^2$

5. The volume of the prism is 365.20 cm³. Find the value of x.

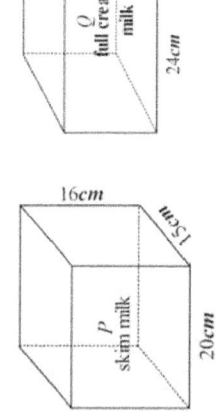

$x = 365.20 \div 22$
$= 16.6\ cm$

CHAPTER 9 : SURFACE AREA AND VOLUME SOLUTIONS

9E APPLICATIONS

1. Sheldon made a bookcase during his woodwork class to keep his comic books. His teacher gave him a 40 cm long, 20 cm wide and 8 cm high rectangular wooden block. He carved out an 18 cm by 4 cm rectangular hole in the block. Calculate the volume of the bookcase.

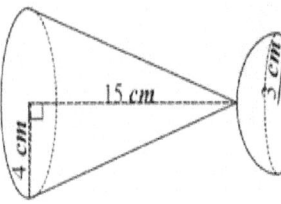

$V = (40 \times 20 \times 8) - (18 \times 20 \times 4)$

$= 4960 \; cm^3$

2. The diagram represents a solid block of wood of length 20m. The faces ABFE, BCGF and ADHE are horizontal rectangles. The faces ABCD and EFGH are vertical rectangles. BC = AD =12m, BF = AE = 8m and FG = EH =18m.

Calculate

(a) the length of CG,

$CG = \sqrt{8^2 + 6^2} = 10 \; m$

(b) the volume of the block,

$Area \; of \; BCFG = \left(\dfrac{18 + 12}{2}\right) \times 8 = 120 \; m^2$

$V = 120 \times 20 = 2400 \; m^3$

(c) the total surface area of the block.

$TSA = 18 \times 20 + 20 \times 8 + 12 \times 20 + 10 \times 20 + 120 + 120 = 1200 \; m^2$

3. The diagram shows a wine glass. The base is a hemisphere of radius 3 cm and the top part is a cone of radius 4cm and vertical height 15 cm.

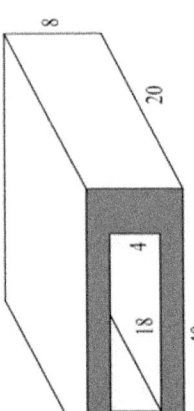

(a) Determine the volume of the whole container.

$Volume = \dfrac{1}{3} \times \pi \times 4^2 \times 15 + \dfrac{2}{3} \times \pi \times 4^3$

$= 385.37 \; cm^3$

(b) Peter poured 280 mL of wine in the glass. Would the wine overflow?

$\dfrac{1}{3} \times \pi \times 4^2 \times h = 280$

$h = \dfrac{280 \times 3}{16\pi} = 16.71 \; cm$

Yes it will overflow.

4. Teddy toad lives in a castle made from mud and sticks under the water. He wants to start a family so he needs to work out how many little taddies he can have. Each one will require a space of 5 cm³. He needs at least 150 cm³ for himself. Calculate how many taddies he can have.

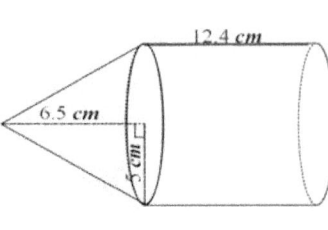

$Volume \; of \; cone = \dfrac{1}{3} \times \pi \times 5^2 \times 6.5 = 170.17 \; cm^3$

$Volume \; of \; cylinder = \pi \times 5^2 \times 12.4 = 973.89 \; cm^3$

$Total \; Volume = 170.17 + 973.89 = 1144.06 \; cm^3$

$Number \; of \; taddies = \dfrac{1144.06 - 150}{5} \approx 199$

CHAPTER 9 : SURFACE AREA AND VOLUME SOLUTIONS

5. A water tanker delivers water in a cylindrical container of length 18 m and radius 1.5 m. After several deliveries, the water remaining in the container is shown in the diagram.

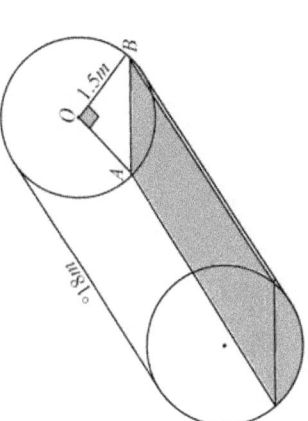

AB is horizontal, O is the centre of the circular cross-section and $\angle AOB = 90°$.

(i) Calculate the area of the minor sector OAB, to three decimal places.

$$A = \frac{90}{360} \times \pi \times 1.5^2 = 1.767\ m^2$$

(ii) Calculate the area of $\triangle$OAB.

$$\text{Area of } \triangle OAB = \frac{1}{2} \times 1.5 \times 1.5 = 1.125\ m^2$$

(iii) Hence calculate the curved surface area of the container that is in contact with the water.

$$\text{Area of sector} - \text{Area of } \triangle = 1.77 - 1.125 = 0.64\ m^2$$

(iv) Calculate the volume of water remaining in the container.

$$V = 0.64 \times 18 = 11.52\ m^3$$

MATHEMATICS APPLICATIONS UNIT 1

6. The diagram on the right shows a garage shed as advertised in the yellow pages. As an investigation, Mr Thompson, the Mathematics teacher, has assigned his students to work out

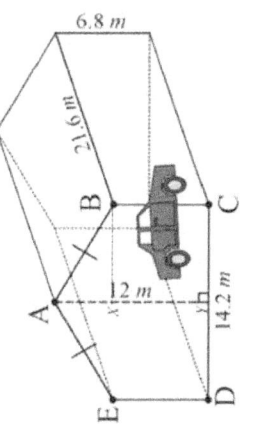

(a) the length of AB.

$AX = 12 - 6.8 = 5.2m$
$BX = 14.2 \div 2 = 7.1m$
$AB = \sqrt{5.2^2 + 7.1^2} = 8.8\ m$

(b) Area of the face ABCDE,

$$\text{Area of trapezium } ABCY = \frac{12+6.8}{2} \times 7.1 = 66.74\ m^2$$
$$\text{Area of } ABCDE = 66.74 \times 2 = 133.48\ m^2$$

(c) the volume of the shed.

$$Volume = 133.48 \times 21.6 = 2883.17\ m^3$$

(d) the total surface area of the shed.

$TSA = 133.48 \times 2 + 21.6 \times 6.8 \times 2 + 8.8 \times 21.6 \times 2 + 14.2 \times 21.6$
$= 1247.60 m^2$

CHAPTER 10

SIMILARITY

10A SIMILAR FIGURES AND SIMILAR TRIANGLES

Similar figures have identical shape, but differ in size. Similar figures are different from congruent figures. The latter have the same shape and are exactly the same size.
The calculators shown below are similar. Clearly, the length and width of the smaller calculator has been multiplied by the same scale factor of 2 to produce the length and the width of the larger calculator.

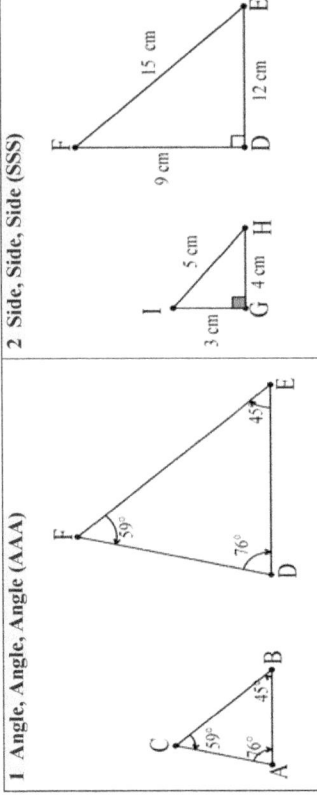

Consider the triangles below. They are obviously similar.

Use the angles to match each corresponding sides.

$\dfrac{EF}{CB} = \dfrac{6}{4} = 1.5$

$\dfrac{DF}{AC} = \dfrac{9}{6} = 1.5$

$\dfrac{DE}{AB} = \dfrac{4.5}{3} = 1.5$

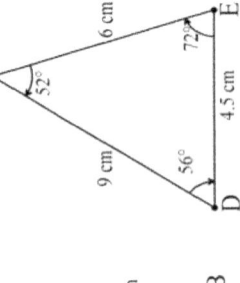

In the triangles ABC and DEF, matching angles are equal as well as the sides have equal ratios (scale factor). Hence, we can say that △ ABC is similar to △ DEF.
Symbolically, we write △ ABC ~ △ DEF.

As a conclusion, in similar figures

- All corresponding sides are in the same ratio
- All pairs of corresponding angles are equal.

10B TESTING SIMILAR TRIANGLES

To prove two triangles are similar, we do not need to know all the sides and all the angles. The four tests listed below can be used to show that two triangles are similar. Depending on what information is given, we have to choose the appropriate test.

1 Angle, Angle, Angle (AAA)

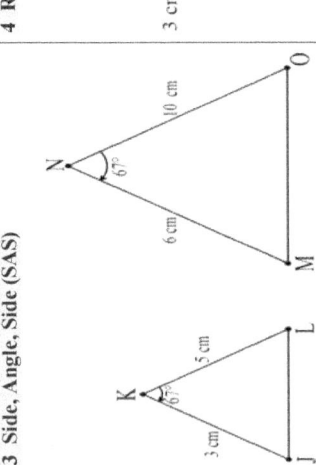

Since three pairs of angles are of the same magnitude,
△ ABC ~ △ DEF (AAA).
If any two angles are given in a triangle we can work out the third angle as angle sum of a triangle is 180°.

2 Side, Side, Side (SSS)

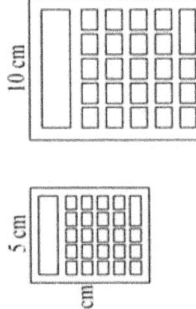

Since three pairs of matching sides are in the same ratio, scale factor being 3.
△ GHI ~ △ DEF (SSS).

3 Side, Angle, Side (SAS)

Since two pairs of corresponding sides are in the same ratio and the angle included between the two sides are equal, we say
△ JKL ~ △ MNO (SAS)

4 Right Angle, Hypotenuse, Side (RHS)

Since the hypotenuse and one side in △PQR is in the same ratio as the hypotenuse and one side of △STU, we can say that
△ PQR ~ △ STU (RHS)

CHAPTER 10 : SIMILARITY SOLUTIONS

EXAMPLE

Use similarity test to show that the following triangles are similar.

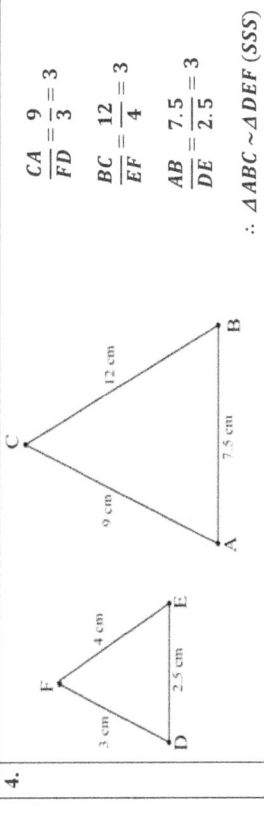

$\angle CAB = 180 - 124 - 19 = 37°$

$\angle DFE = 180 - 124 - 37 = 19°$

$\angle A = \angle D = 37°,$
$\angle B = \angle E = 124°$
$\angle C = \angle F = 19°$
$\therefore \triangle ABC \sim \triangle DEF \ (AAA)$

EXERCISE 10B

Use similarity test to show that the following triangles are similar.

1.	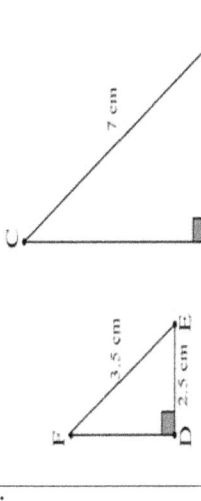	$\angle A = \angle D = 72°,$ $\angle B = \angle E = 56°,$ $\angle C = \angle F = 52°$ $\therefore \triangle ABC \sim \triangle DEF \ (AAA)$
2.		$\angle CAB = 180 - 108 - 27 = 45°$ $\angle DEF = 180 - 45 - 27 = 108°$ $\angle A = \angle D = 45°,$ $\angle B = \angle E = 108°,$ $\angle C = \angle F = 27°$ $\therefore \triangle ABC \sim \triangle DEF \ (AAA)$
3.		$\dfrac{ED}{CA} = \dfrac{5}{2} = 2.5$ $\dfrac{EF}{CB} = \dfrac{10}{4} = 2.5$ $\angle C = \angle F = 53°$ $\therefore \triangle ABC \sim \triangle DEF \ (SAS)$

MATHEMATICS APPLICATIONS UNIT 1

Use similarity test to show that the following triangles are similar.

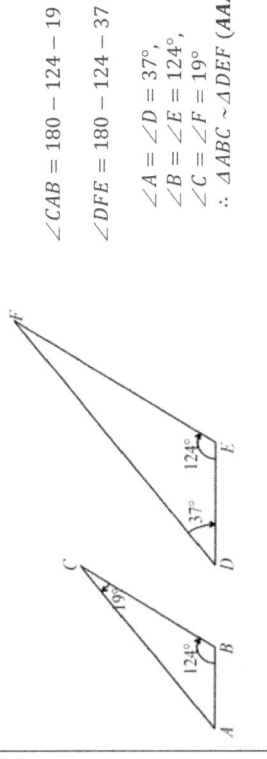
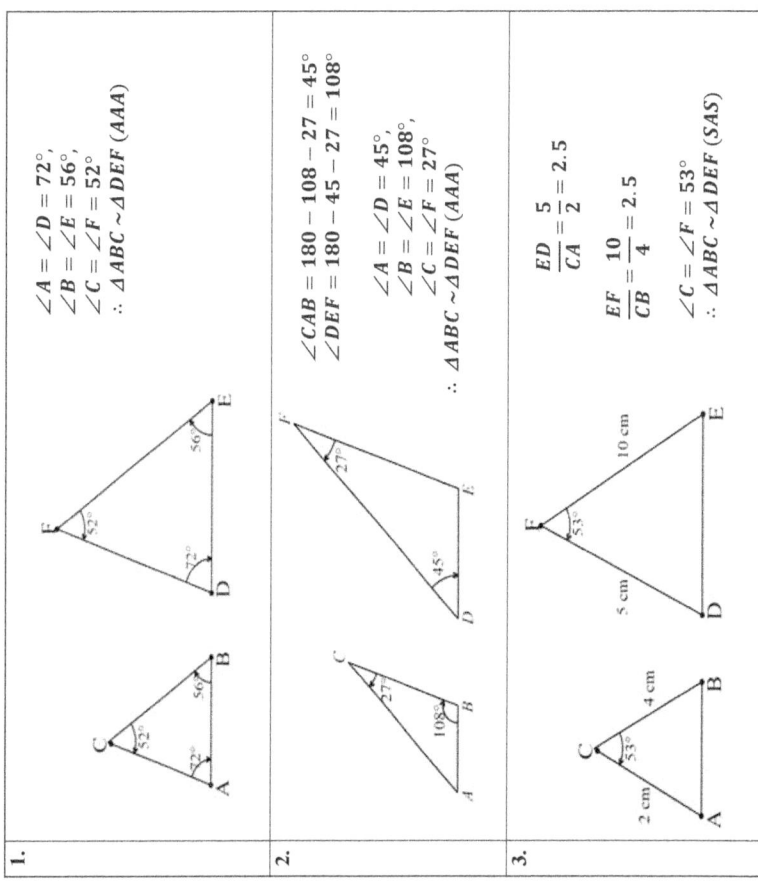

4.		$\dfrac{CA}{FD} = \dfrac{9}{3} = 3$ $\dfrac{BC}{EF} = \dfrac{12}{4} = 3$ $\dfrac{AB}{DE} = \dfrac{7.5}{2.5} = 3$ $\therefore \triangle ABC \sim \triangle DEF \ (SSS)$
5.		$\dfrac{CB}{FE} = \dfrac{7}{3.5} = 2$ $\dfrac{AB}{DE} = \dfrac{5}{2.5} = 2$ $\angle A = \angle D = 90°$ $\therefore \triangle ABC \sim \triangle DEF \ (RHS)$
6.		$\dfrac{CA}{FD} = \dfrac{15}{10} = 1.5$ $\dfrac{BC}{EF} = \dfrac{18}{12} = 1.5$ $\angle C = \angle F = 44°$ $\therefore \triangle ABC \sim \triangle DEF \ (SAS)$
7.		$\dfrac{CA}{FE} = \dfrac{15}{7.5} = 2$ $\dfrac{AB}{DE} = \dfrac{12}{6} = 2$ $\angle B = \angle D = 90°$ $\therefore \triangle ABC \sim \triangle DEF \ (RHS)$

CHAPTER 10 : SIMILARITY SOLUTIONS

10C SIMILARITY AND SCALE FACTOR

If two triangles are similar, the scale factor is the ratio of the corresponding sides.

In the triangles on the right,

- The side corresponding to RQ is UT
- The side corresponding to PR is SU

$$\frac{PR}{SU} = \frac{RQ}{UT} = \frac{1}{2}$$

∴ $\triangle PQR$ is half the size of $\triangle STU$ in terms of length.

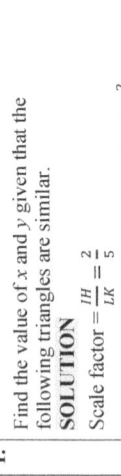

EXAMPLES

1. Find the value of x and y given that the following triangles are similar.

   SOLUTION

   Scale factor = $\frac{IH}{LK} = \frac{2}{5}$

   To find y we multiply the size of LJ by $\frac{2}{5}$

   ∴ $y = 12 \times \frac{2}{5} = 4.8$

   To find x however, we need to multiply by the reciprocal of the scale factor $\frac{5}{2}$

   ∴ $x = 4 \times \frac{5}{2} = 10\ cm$

   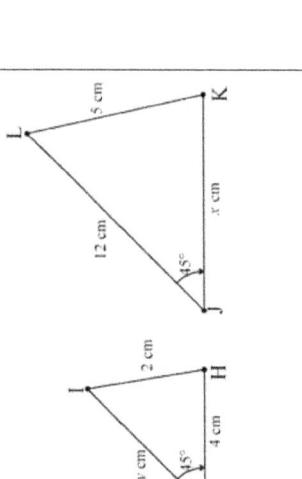

2. In the diagram BD is parallel to CE.
   Given that $\triangle ABD \sim \triangle ACE$, find the value of x.

   SOLUTION

   Scale factor = $\frac{AB}{AC} = \frac{3}{5}$

   BD corresponds to the side CE

   To find x, we have to multiply 10 by $\frac{3}{5}$

   $x = 10 \times \frac{3}{5} = 6\ cm$.

   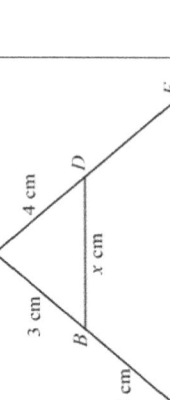

EXERCISE 10C

Use similar triangles to find the value of the pronumerals in each case.

1. Scale factor = $\frac{FD}{CA} = \frac{5}{2} = 2.5$

   To find x we multiply the size of BC by 2.5

   ∴ $x = 4 \times 2.5 = 10\ cm$

   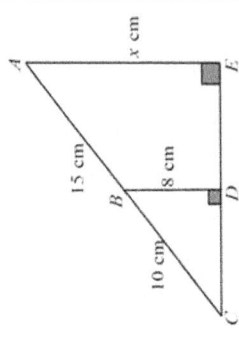

2. Scale factor = $\frac{AC}{BC} = \frac{25}{10} = 2.5$

   To find x we multiply the size of BD by 2.5

   $x = 8 \times 2.5 = 20\ cm$

   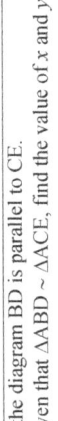

3. In the diagram BD is parallel to CE.
   Given that $\triangle ABD \sim \triangle ACE$, find the value of x and y.

   Scale factor = $\frac{AB}{AC} = \frac{6}{8} = 0.75$

   To find x we multiply the size of CE by 0.75

   ∴ $x = 12 \times 0.75 = 9\ cm$

   $Solve\ \left(\frac{y}{y+3} = 0.75\right)$

   ∴ $y = 9$

   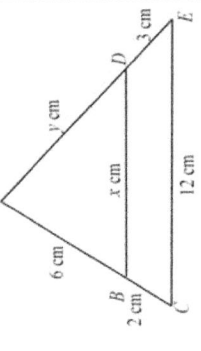

4. In the diagram BD is parallel to AE.
   Given that $\triangle BCD \sim \triangle ACE$, find the value of x and y.

   Scale factor = $\frac{AE}{BD} = \frac{6}{4} = 1.5$

   To find x we multiply the size of BC by 1.5

   $AC = 3 \times 1.5 = 4.5\ cm$

   $x = 4.5 - 3 = 1.5\ cm$

   $Solve\ (1.5y = y + 2)$ ∴ $y = 4$

   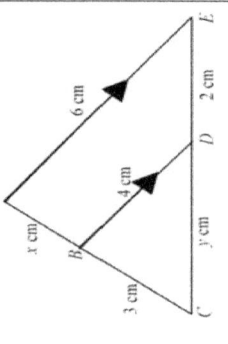

5. AOB and COD are straight lines.

(a) Show that triangles OCA and ODB are similar.

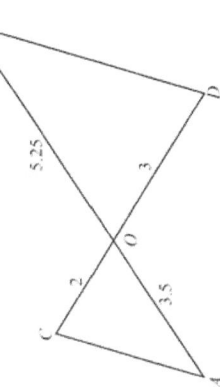

$$\frac{OD}{OC} = \frac{3}{2} = 1.5$$

$$\frac{OB}{OA} = \frac{5.25}{3.5} = 1.5$$

∠COA = ∠BOD (vertically opp.)

∴ △OCA ~ △ODB (SAS)

(b) Given that BD = 3.9 cm, find AC.

$$AC = \frac{3.9}{1.5} = 2.6\ cm$$

6. In the given diagram, AC is parallel to DB.

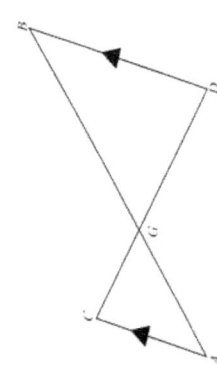

At each step of proof below, the statement and some reasons are provided. Complete the table below.

step	Statement	Reason
1	∠CGA = ∠BGD	Vertically opposite angles
2	∠ACG = ∠BDG	Alternate angles
3	∠CAG = ∠DBG	Alternate angles
4	△ CAG is similar to △ BDG	AAA, SAS, SSS, RHS (choose one)

7. Show that the two triangles are similar and then find the value of x.

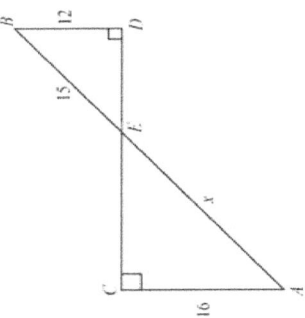

∠CAE = ∠DBE (Alternate angles)
∠CEA = ∠BED (Vertically opposite angles)
∠ACE = ∠BDE (GIVEN)
∴ △ACE ~ △BED (AAA)

$$Scale\ factor = \frac{16}{12} = \frac{4}{3}$$

$$\therefore x = \frac{4}{3} \times 15 = 20$$

8. In the diagram, which is not drawn to scale, ∠ADB = ∠ACD = 37° and AD = DC.

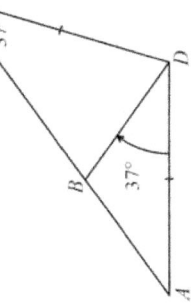

(a) Write down a second pair of equal angles.

∠ABD = ∠ADC

(b) Complete the two statements below:

$$\frac{AC}{AD} = \frac{AD}{AB}$$

$$\frac{\triangle ABD}{\triangle ADC} = \frac{AD^2}{AB^2}$$

9. In the triangle ABC, AB = 6 cm, AC = 9 cm and D is the point on the side AC such that ∠ABD = ∠ACB = 40°.

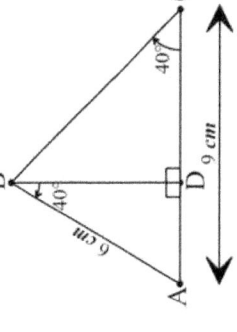

(a) Write down another pair of equal angles.

∠ABC = ∠ADB

(b) Use similar triangles to find the length of AD.

$$\frac{6}{AD} = \frac{9}{6}$$

$$AD = 4\ cm$$

CHAPTER 10 : SIMILARITY SOLUTIONS

10D   AREA OF SIMILAR FIGURES

If two polygons are **similar**, the ratio of their **areas** is equal to the **square** of the ratio of their corresponding sides.

In general for two similar figures having areas A_1 and A_2 and corresponding sides l_1 and l_2 respectively, $\frac{A_1}{A_2} = \left(\frac{l_1}{l_2}\right)^2$ or $A_1 = A_2 \times \left(\frac{l_1}{l_2}\right)^2$ or $A_2 = A_1 \times \left(\frac{l_2}{l_1}\right)^2$ or $\frac{l_1}{l_2} = \sqrt{\frac{A_1}{A_2}}$

EXAMPLE 1
The diagram shows two shapes. Shape 1 has an area of 12 cm². Given that the two shapes are similar, find the area of shape 2.

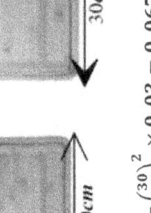

shape 1 — $A = 12\ cm^2$, 3cm
shape 2 — $A = ?$, 6 cm

SOLUTION
Area of shape 2 = (Scale factor)² × Area of shape 1
$= \left(\frac{6}{3}\right)^2 \times 12$
$= 48\ cm^2$

EXAMPLE 2
The diagram shows two similar kites. The area of the larger kite is 40 cm². Find the area of the smaller kite.

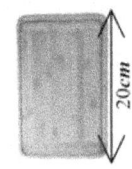

$A = ?$, 3 ; $A = 40\ cm^2$, 5

SOLUTION
Area of smaller kite $= \left(\frac{3}{5}\right)^2 \times 40$
$= 14.4\ cm^2$

EXAMPLE 3
Two similar figures have areas 12 cm² and 27 cm² respectively. Find the ratio of their corresponding sides.

SOLUTION
Using $\frac{l_1}{l_2} = \sqrt{\frac{A_1}{A_2}}$

$\frac{l_1}{l_2} = \sqrt{\frac{12}{27}} = \sqrt{\frac{4}{9}} = \frac{2}{3}$

The ratio of their corresponding sides is 2 : 3.

EXAMPLE 4
The diagram shows two similar regular heptagons. Find the value of x.

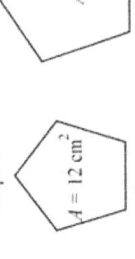

27 cm², 3 cm ; 48 cm², x cm

SOLUTION
Using $\frac{l_1}{l_2} = \sqrt{\frac{A_1}{A_2}}$

$\frac{3}{x} = \sqrt{\frac{27}{48}}$

$\frac{3}{x} = \sqrt{\frac{9}{16}} = \frac{3}{4}$

$\therefore x = 4$

EXERCISE 10D

1. A photo frame and its enlargement are similar in shape. The smaller photograph has an area of 42 cm². Calculate the area of the larger photograph.

   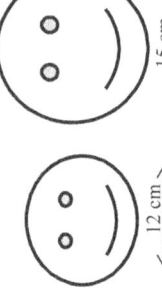

   10cm ; 42cm² ; 15cm

   $A = \left(\frac{15}{10}\right)^2 \times 42 = 94.5\ cm^2$

2. Two chopping boards are similar in shape. The smaller chopping board is 20 cm long and has an area of 0.03 m². Calculate the area of the larger chopping board.

   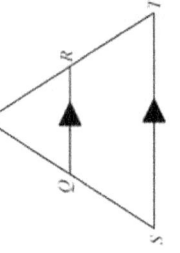

   20cm ; 30cm

   $A = \left(\frac{30}{20}\right)^2 \times 0.03 = 0.0675\ m^2$

3. Two regular heptagons are mathematically similar in shape. The larger heptagon has an area of 4600 mm². Find the area of the smaller heptagon.

   30 mm ; 4600 mm² ; 50 mm

   $A = \left(\frac{30}{50}\right)^2 \times 4600 = 1656\ mm^2$

4. Two smiley faces shaped symbols are similar in shape. The smaller shape has an area of 113 cm². Calculate the area of the larger shape.

   12 cm ; 15 cm

   $A = \left(\frac{15}{12}\right)^2 \times 113 = 176.56\ cm^2$

5. In the diagram, QR is parallel to ST and $\frac{PQ}{PS} = \frac{3}{8}$.

   (a) Find the value of $\frac{Area\ of\ \Delta PQR}{Area\ of\ \Delta PST} = \frac{9}{64}$.

   $\sqrt{\frac{9}{64}} = \frac{3}{8}$

   (b) Given that the area of the triangle PQR is 36 cm², find the area of the trapezium QRST.

   $Area\ of\ \Delta PST = \left(\frac{8}{3}\right)^2 \times 36 = 256$

   $\therefore Area\ of\ QRST = 256 - 36 = 220\ cm^2$.

CHAPTER 10 : SIMILARITY SOLUTIONS

6. A rectangle has an area of 15 cm². Find the area of a similar rectangle whose dimensions are twice of the given rectangle.

$$A = 2^2 \times 15 = 60 \; cm^2$$

7. A square has an area of 60 cm². Find the area of a square having three times the length.

$$A = 3^2 \times 60 = 540 \; cm^2$$

8. A circle has an area of 12 cm². Find the area of a circle with radius half of the length of the given circle.

$$A = 0.5^2 \times 12 = 3 \; cm^2$$

9. Find the ratio of the corresponding sides of two similar triangles if their areas are 16 cm² and 64 cm².

$$\sqrt{\frac{16}{64}} = \frac{1}{2} \; or \; 1:2$$

10. The following triangles are similar. Find the area of the triangle marked A.

$$A = \left(\frac{3}{2}\right)^2 \times 24 = 54 \; cm^2$$

11. The following triangles are similar. Find the area of the triangle marked A.

$$A = \left(\frac{3}{5}\right)^2 \times 27 = 9.72 \; cm^2$$

12. A floor is covered by 600 tiles which are 10 cm by 10 cm. How many 20 cm by 20 cm tiles are needed to cover the same floor?

$$600 \div 2^2 = 150 \; tiles$$

13. A photo is 10 cm long. It is enlarged so that all dimensions are increased by 20%. Find the ratio of the area of the enlarged photo to the area of the original photo.

$$1.2^2 : 1^2 = 1.44 : 1$$

10E   VOLUME OF SIMILAR FIGURES

If two solids are **geometrically similar**, the ratio of their **volumes** is equal to the **cube** of the ratio of their corresponding sides or radius.

In general for two similar figures having volumes V_1 and V_2 and corresponding sides l_1 and l_2 respectively, $\frac{V_1}{V_2} = \left(\frac{l_1}{l_2}\right)^3$ or $V_2 = V_1 \times \left(\frac{l_2}{l_1}\right)^3$ or $\frac{l_1}{l_2} = \sqrt[3]{\frac{V_1}{V_2}}$

EXAMPLES

1. Two similar cylinders have radii 2 cm and 5 cm respectively. Find the ratio of their volumes.

SOLUTION

$$\frac{V_1}{V_2} = \left(\frac{r_1}{r_2}\right)^3$$
$$= \left(\frac{2}{5}\right)^3$$
$$= \frac{8}{125} \; or \; 8:125$$

2. Two similar cones have heights 5 cm and 10 cm respectively. If the volume of the smaller cone is 25 cm³, find the volume of the larger cone.

SOLUTION

Scale factor $= \frac{10}{5} = 2$

Volume of larger cone $= 2^3 \times 25 = 200 \; cm^3$

3. Two solid spheres have volume 80 cm³ and 270 cm³. Find the ratio of their radii.

SOLUTION

Using

$$\frac{l_1}{l_2} = \sqrt[3]{\frac{V_1}{V_2}} = \sqrt[3]{\frac{80}{270}}$$
$$= \sqrt[3]{\frac{8}{27}}$$
$$= \frac{2}{3} \; or \; 2:3$$

4. Two cylindrical containers have volumes 16 cm³ and 54 cm³ respectively. If the radius of the smaller cylinder is 6 cm, find the radius of the larger cylinder.

SOLUTION

Scale factor $= \sqrt[3]{\frac{V_1}{V_2}} = \sqrt[3]{\frac{54}{16}}$
$$= \sqrt[3]{\frac{27}{8}}$$
$$= \frac{3}{2}$$

$radius = \frac{3}{2} \times 6 = 9 \; cm$

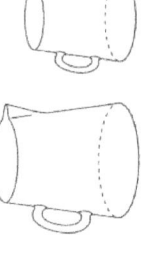

5. Two geometrically similar jugs have volumes of 1000 cm³ and 512 cm³. They have circular bases. The diameter of the base of the larger jug is 9 cm. Calculate the diameter of the base of the smaller jug.

SOLUTION

Scale factor $= \sqrt[3]{\frac{V_1}{V_2}} = \sqrt[3]{\frac{512}{1000}} = \frac{4}{5}$

Diameter of smaller jug $= \frac{4}{5} \times 9 = 7.2 \; cm.$

EXERCISE 10E

1. Two similar cylinders have radii 3 cm and 5 cm respectively. Find the ratio of their volumes.

$$ratio = \left(\frac{3}{5}\right)^3 = \frac{27}{125}$$

or 27 : 125

2. Two spheres have radii 4 cm and 7 cm respectively, find the ratio of their volumes.

$$ratio = \left(\frac{4}{7}\right)^3 = \frac{64}{343}$$

or 64 : 343

3. Two similar cylindrical jugs have volume 6 cm³ and 48 cm³. Find the ratio of their radii.

$$ratio = \sqrt[3]{\frac{6}{48}} = \sqrt[3]{\frac{1}{8}} = \frac{1}{2}$$

or 1 : 2

4. Two similar cylinders have heights 3 cm and 6 cm respectively. If the volume of the smaller cylinder is 36 cm³, find the volume of the larger cylinder.

Scale factor $= \frac{6}{3} = 2$
Volume of larger cylinder
$= 2^3 \times 36$
$= 288\ cm^3$

5. Two similar solids have heights 2 cm and 5 cm respectively. If the volume of the larger solid is 1000 cm³, find the volume of the smaller solid.

Scale factor $= \frac{2}{5}$
Volume of smaller solid
$= \left(\frac{2}{5}\right)^3 \times 1000 = 64\ cm^3$

6. Two similar jugs have volumes 200 cm³ and 25 cm³ respectively. If the radius of the smaller jug is 3 cm, find the radius of the larger jug.

Scale factor $= \sqrt[3]{\frac{200}{25}} = 2$
$\therefore$ **radius of larger jug** $= 2 \times 3$
$= 6\ cm$

7. Given that the two solids are similar, find the value of V.

Scale factor $= \frac{10}{5} = 2$
Volume $= 2^3 \times 60 = 480\ cm^3$

8. Given that the two cylinders are similar, find the value of V.

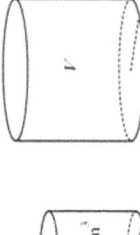

Scale factor $= \frac{9}{6} = 1.5$
Volume $= 1.5^3 \times 24 = 81\ cm^3$

9. A cylinder K has a volume of 200 cm³. Calculate the volume of a cylinder similar to K but with radius twice that of K.

Volume of larger cylinder $= 2^3 \times 200$
$= 1600\ cm^3$

10. Mary makes two geometrically similar cakes.

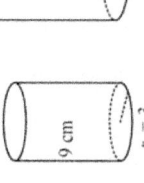

The heights of the cakes are 6 cm and 9 cm.

Mary uses 1200 cm³ of cake mixture to make the smaller cake.
Find the volume of cake mixture she uses to make the larger cake.

Scale factor $= \frac{9}{6} = 1.5$
Volume of larger cake
$= 1.5^3 \times 1200 = 4050\ cm^3$

11. Two cylinders are geometrically similar. The radius of the smaller one is 3 cm. The radius of the larger one is 6 cm.

(a) The height of the smaller cylinder is 9 cm. Find the height of the larger cylinder.

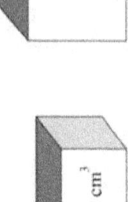

Scale factor $= \frac{6}{3} = 2$
Height of larger cylinder $= 2 \times 9 = 18\ cm$

(b) Find the ratio of the volume of the larger cylinder to the surface area of the smaller.

Volume of larger cylinder $= \pi \times 6^2 \times 18 = 2035.75\ cm^3$
Surface area of smaller cylinder $= 2\pi \times 3^2 + 2\pi \times 3 \times 9 = 226.19$
Ratio $= 2035.75 : 226.19$
$\approx 9 : 1$

CHAPTER 10 : SIMILARITY SOLUTIONS

12. Tracy buys two tins of baked beans in Coles. The tins are geometrically similar to each other. The height of one of the tin is 8 cm and the height of the other is 12 cm.

(a) The radius of the smaller tin is 5 cm. Calculate the radius of the larger tin.

Scale factor $= \frac{12}{8} = 1.5$

Radius of larger tin $= 1.5 \times 5 = 7.5 cm$

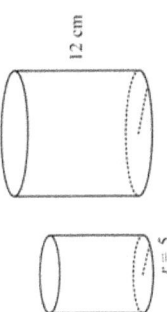

(b) Calculate the ratio of the volume of the smaller tin to the volume of the larger tin.

$$ratio = \left(\frac{8}{12}\right)^3 = \left(\frac{2}{3}\right)^3 = \frac{8}{27}$$

or 8 : 27

(c) The cost of a small tin of beans is $2.40. Calculate the cost of a large tin of beans assuming that there is no reduction for buying the larger tin.

$$Cost = \frac{27}{8} \times 2.40 = \$8.10$$

13. The two containers shown in the diagram are geometrically similar. Their heights are 30 cm and 40 cm as shown.

(a) The radius of the smaller container is 9 cm. Calculate the radius of the base of the larger container.

Scale factor $= \frac{40}{30} = \frac{4}{3}$

Radius of larger container $= \frac{4}{3} \times 9 = 12cm$

(b) The containers are completely filled with water. Given that the larger container holds 32 litres of water, calculate the capacity of the smaller container.

$$Capacity\ of\ smaller\ container = \left(\frac{3}{4}\right)^3 \times 32 = 13.5\ litres.$$

MATHEMATICS APPLICATIONS UNIT 1

10 F MAPS, SCALE DRAWINGS AND BUILDING PLANS

Maps are usually smaller than real things as it is not possible to draw a map the same size as the actual area. However, angles and directions on maps are similar to real life situations. The only difference between a map and reality is its size and we use a scale to depict this difference.

$$map\ scale = \frac{map\ length}{Actual\ length}$$

Map scales are usually written in the form 1 : n, meaning 1 cm on the map represents n cm in actual life. For example, a scale of 1 : 20 000 implies 1 cm on the map represents 20 000 cm in reality.

CLASS ACTIVITY

Write the following scales in the form 1 : n. The first 2 questions have been done as examples.

1. 1 cm represents 2 m	2. 1cm represents 3 km	3. 5 cm represents 2 km
$1\ cm \rightarrow 2\ m$ $1\ cm \rightarrow 200\ cm$ *scale is* 1:200	$1\ cm \rightarrow 3\ km$ $1\ cm \rightarrow 3000 \times 100\ cm$ *scale is* 1:300 000	$5\ cm \rightarrow 2\ km$ $5\ cm \rightarrow 2000 \times 100\ cm$ $1\ cm \rightarrow 40\ 000\ cm$ *scale is* **1 : 40 000**
4. 1 cm represents 5 m	5. 1cm represents 5 km	6. 2 cm represents 8 km
$\mathbf{1\ cm \rightarrow 5\ m}$ $\mathbf{1\ cm \rightarrow 500\ cm}$ ***scale is* 1 : 500**	$\mathbf{1\ cm \rightarrow 5\ km}$ $\mathbf{1\ cm \rightarrow 5000 \times 100\ cm}$ ***scale is* 1 : 500 000**	$\mathbf{2\ cm \rightarrow 8\ km}$ $\mathbf{2\ cm \rightarrow 8000 \times 100\ cm}$ $\mathbf{1\ cm \rightarrow 400\ 000\ cm}$ ***scale is* 1 : 400 000**

7. The diagram below shows a scale drawing of the Murray Darling Bridge in South Mississippi.

By making suitable measurements, determine

(a) the height of the bridge, $\approx 4\ cm \rightarrow 4 \times 4 = 16m$

(b) Santa's Christmas truck measures 4.5m high. Can it make it through the bridge?

$\approx 1.1\ cm \rightarrow 1.1 \times 4 = 4.4m$ ∴ *No*

CHAPTER 10 : SIMILARITY SOLUTIONS

8. The diagram shows the front elevation of Beach Side Apartments in North Carolina. Using the scale shown, determine

BEACH SIDE APARTMENTS

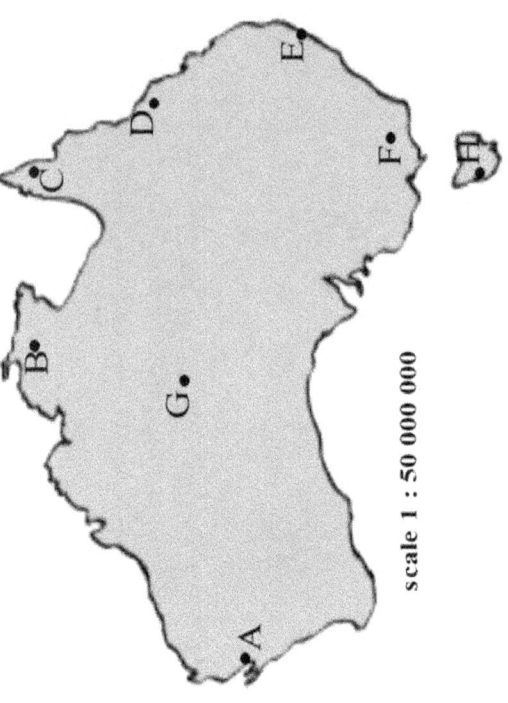

(a) the actual height of the building

$scale\ 1.2cm \rightarrow 10m$
$height \approx 6.1\ cm$
$Actual\ height = \frac{6.1}{1.2} \times 10$
$= 50.83m$

(b) the actual width of the building.

$width \approx 4.8cm$
$Actual\ width = \frac{4.8}{1.2} \times 10$
$= 40m$

9. The diagram below shows the plan of a house in Perth City. The scale of the drawing is 1 : 100. (1cm represents 100 cm)

DREAM HOUSE CORPORATION

By making suitable measurements, determine

(a) The area of BED 3.

$2 \times 2.9 = 2m \times 2.9m = 5.8\ m^2$

(b) The difference in area between the garage and the alfresco area.

$(6.2 \times 2.9) - (4.5 \times 2.2) = 9.9\ m^2$

MATHEMATICS APPLICATIONS UNIT 1

10. The diagram shows a scale drawing of a newly found country. The scale is 1 : 50 000 000.

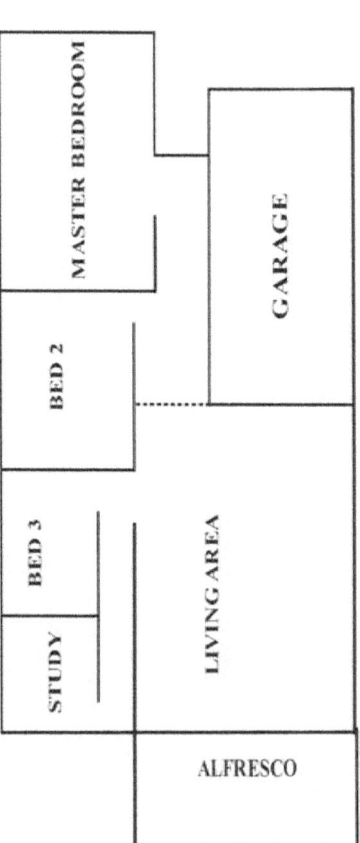

scale 1 : 50 000 000

(a) Determine how many kilometres does 1 cm on the map represent in actual.

$50\ 000\ 000 \div 100 \div 1000 = 500\ km$
$\therefore 1cm \rightarrow 500\ km$

(b) Using appropriate measurements, state the distance in kilometres between A and D.

$\approx 11\ cm \rightarrow 11 \times 500 = 5500\ km$

(c) By how many kilometres is F farther from G compared to B?

$G\ to\ F \approx 6.5\ cm \rightarrow 3250\ km$
$G\ to\ B \approx 3\ cm \rightarrow 1500\ km$
$\therefore 3250 - 1500 = 1750\ km$

www.ingramcontent.com/pod-product-compliance
Lightning Source LLC
Chambersburg PA
CBHW080236250426
43670CB00043BA/2562